Advanced Mathematics
for
FPGA and DSP Programmers

Advanced Mathematics
for
FPGA and DSP Programmers

Conquering Fixed-Point Pitfalls

Tim Cooper for Amches Inc.

Leapin Leo Press
Brevard, NC
www.leapinleo.com

Printed in the United States of America.

Cover art by Taryn Delinsky.
Typesetting by Mountain River Communications, LLC.

ISBN 978-0-9790581-1-0

Table of Contents

Table of Figures

Introduction

Congratulations on your purchase of Advanced Mathematics for FPGA and DSP Programmers. This text has been taught by author Tim Cooper to his clients with great success. Topics range from the importance of correct choices for numerical representation to subjects like correlation and convolution.

Scattered throughout the book are Amches Pointers that are similar to the explanatory sidebars you find in many books:

Amches Pointer

Each additional bit of width provides 6 dB more dynamic range, and also a 6 dB improvement in quantization noise. Audio CD players have dynamic range of 96 dB and SNR of 96 dB, suggesting that the data samples are 16 bits wide.

Graphs are used throughout the text as well (see examples below). See the list of figures before this introduction.

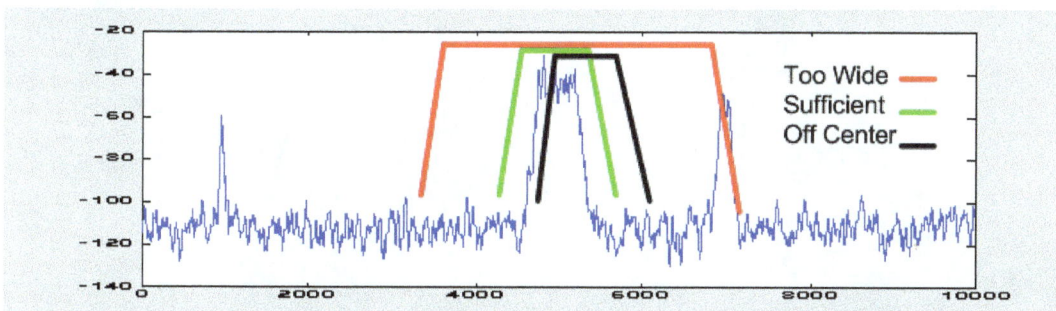

Figure 13. Signal and Filter Bandwidth matching.

For quicker help, feel free to go directly to a chapter that addresses a problem you might be facing, and then work your way through the rest of the material later.

Happy problem solving!

The Amches Publishing Team

www.Amches.com

About the author

Tim Cooper has been developing real-time embedded and signal processing software for commercial and military applications for over 30 years. Mr. Cooper has authored numerous device drivers, board support packages, and signal processing applications for real-time-operating systems.

Mr. Cooper has also authored high-performance signal processing libraries based on SIMD architectures. Other signal processing experience includes MATLAB algorithm development and verification, and working with FPGA engineers to implement and validate signal processing algorithms in VHDL.

Much of Mr. Cooper's experience involves software development for systems having hard real-time requirements and deeply embedded processors, where software reliability, performance, and latency are significant cost drivers. Such systems typically require innovative embedded instrumentation that collects performance data without competing for processing resources.

Mr. Cooper holds a Bachelor of Science in Computer Sciences and a Master's degree in Computer and Electronics Engineering from George Mason University.

Numbers and Representation

1

Software developers choose between floating point, double precision, or long/short/byte data representations. There is little need to be concerned about overflow, underflow, or dynamic range with floating point or double precision. When developing FPGA applications, the developer must choose the width for variables, the signed representation, the location of the radix point, and what actions to take for overflow – representing several hundred choices for numerical representations.

An incorrect choice within an FPGA implementation can result in a system which does not work. Worse, it can result in a system that works "most of the time" – producing useful results until an overflow or some other condition deep within the implementation produces an invalid result.

Choosing the best representation for numerical data within an FPGA is the first critical step in producing a working implementation.

1.1 Representation Choices

When a numerical algorithm is implemented within a processor or FPGA, numbers are represented as a string of binary digits. There are several ways to represent binary numbers. Each representation offers some advantages and disadvantages.

- *Width (# bits):* A processor may restrict the width of a number to 8, 16, or 32 bits. Some processors support 64 bit values. However, within an FPGA, the width of a value can be any integer value, and is fixed by the designer, for each variable and operation, at design time.

- *Fixed* or *Floating Point:* Floating point values require more storage space, more implementation complexity, and therefore more time per operation. However, floating point values provide tremendous dynamic range, and free a developer from worrying about overflow or loss of significance. Because of the complexity, increased resources, and increased time, floating point is generally not used within FPGA implementations.

- *Integer, fraction, or other fixed point representations:* Just as decimal numbers may be integers, fractions, or fixed point quantities having a limited number of digits to the left and right of the decimal point, binary numbers may also be integers, fractions, or have a limited number of bits to the left and right of the radix point.

- *Real or Complex values:* complex numbers are very useful in signal processing applications. A complex number is a pair of numbers that travel together: a real component and an imaginary component. Specific rules dictate how to add, subtract, multiply, divide, and even compare and exponentiate complex numbers. The real and imaginary components may be fixed point, integer, fractional, or floating point values.

- *Sign-magnitude:* numbers use one bit within a value to indicate a negative number. The other bits indicate the magnitude of the number. In this case, the sign bit is similar to a minus sign in standard notation. Signed-magnitude values, though simple, are generally not used.

- *Two's Complement:* Two's complement representation is the most common form used for representing signed values. Like signed-magnitude numbers, one can examine the most significant bit to determine if a value is negative. Unlike signed-magnitude numbers, if the value is negative, the remaining bits are the complement of the value rather than the magnitude of the value.

- *Unsigned Numbers:* Unsigned numbers do not have a sign bit. The most significant bit within an unsigned value is therefore used as the highest-order binary digit of the value.

2

Unsigned numbers are useful for representing quantities that can only take on values greater than or equal to zero, like event counters, addresses, or array indices.

1.2 Two's Complement versus Sign Magnitude

Two's complement notation is preferred over sign magnitude for reasons of implementation simplicity. Consider addition: when adding two sign-magnitude values, an adder must first examine each input operand. If an operand's sign bit is set, the adder must complement the value before adding. After adding, the adder must examine the result to see if it is negative. If it is negative, this post-processing stage must complement it (to determine the magnitude), and set the resulting sign bit.

For two's complement representation, and adder is simply an adder; no pre- or post-processing steps are required.

The number line for of two's complement values differs from that for sign magnitude. Consider an 8 bit value. The two's complement number line for 8 bit values runs from -128 to +127. An 8 bit sign magnitude value lives in the range from -127 and +127, inclusive. An oddity of sign-magnitude representation is that it can represent the value "negative zero" as well as zero.

Two's complement numbers do not support a negative zero; instead they provide one more negative value than positive values. It is possible to represent -128 using 8-bit values, but it is not possible to represent +128. Converting the "largest" negative value to a positive value will result in overflow.

> ### 🌐 *Amches Pointer*
>
> In two's complement, a negative value is "smaller" if it is larger in an unsigned sense. For example, using 16 bit quantities, the number 0xffff is -1, while the number 0x8000 is -32768; but when interpreted as unsigned values, 0xffff is larger than 0x8000.

1.3 Integer versus Fractional Representation

It is common to think of fixed point values as integers: an 8 bit A/D converter may generate values of -128 to +127. However, sometimes it is more convenient to think of these values as fractions between -1 and 1. In this case, for 8 bit values, we divide the integer value by 128 to give the fractional range: -128/128 to 127/128, which is equal to -1.0000 to +0.9921875.

Thinking of these values as fractions does not affect addition, subtraction, or comparison operators, provided all of these operators agree on the location of the radix point.

Fractional representation provides an advantage when multiplying: when multiplying two integers, extra bits must be provided on the left to prevent overflow. But when multiplying two fractional values, the result will always be less than or equal to 1.0; therefore, no extra bits on the left must be provided. Bits on the right can be discarded (with or without rounding), to limit the growth of the result to the desired precision.[1]

The value below is a two's complement signed fractional value, where S is the sign bit. The radix point is between S and the most significant digit. If the sign bit is 0, the value is binary $0.1011011/10000000$, which is $+91/128$.

S	1	0	1	1	0	1	1

1.4 Width and Dynamic Range

The width of a value determines the dynamic range that the value can accommodate. The dynamic range is the ratio of the largest signal that can be represented to the smallest. For an 8-bit unsigned value, the dynamic range is 255 to 1. Normally the dynamic range is specified in power dB. Each bit provides 6 dB of dynamic range, so an 8-bit quantity provides a dynamic range of 48 dB. A signed 8 bit value also provides 48 dB of dynamic range: the smallest signal may vary between 0 and +1, the largest may vary between -128 and +128; this ratio is still 255:1.

1.5 Width and Quantization Noise

An n bit quantity can only represent 2^n different values. If a time-varying analog signal is fed into an 8-bit A/D converter, the result will be the closest binary value that represents the signal voltage. Because the voltage is continuous, it often takes on values "in between" the values that can be represented by the 8 bit quantity. The converter is said to *quantize* the input voltage level to a fixed number of discrete values.

This slight difference between the actual voltage and the binary representation provided by the ADC can be thought of as a random noise value that is between 0 and ½ bits for each sample from the ADC. This noise is *quantization noise*.[2]

---

[1] An exception: when computing $(-1) \times (-1)$, the result, +1, cannot be represented in fractional form. Fractional multipliers must detect this condition and replace the result with the largest positive value less than one that can be represented in the given number of bits.

[2] Examine the specifications for an ADC. This value is the best theoretical case; ADCs may have slightly worse behavior.

Quantization noise is also introduced into a system when algorithms reduce the least significant bits of a result through truncation or rounding.

Quantization noise can be reduced by using a wider ADC (for example, 10 bits instead of 8) spanning the same voltage range. Within an FPGA, quantization noise can also be reduced by truncating only those low order bits that provide no signal information.

The quantization noise figure of an ADC assumes that the input signal varies over the full range of the ADC input voltage. For example, if an ADC accepts -1V to 1V, and the signal into the ADC operates over this range, then the input signal will span the full range of ADC binary output values.

If, however, the ADC accepts -1V to 1V, but the input signal spans a range of 0.25V to -0.25V, then the ADC will provide binary values that span a smaller range. Quantization noise *for the signal* will be increased (because, in this case we are using ¼ of all possible ADC output values).

A signal that spans ¼ of the ADC input range for an 8 bit converter provide only 6 effective bits of output. The quantization noise for the ADC *itself* is the same (1/256 of the full voltage range), but quantization noise for the signal is 1/64, four times this figure. This shows the importance of matching the input signal level to the range of the ADC.

1.6 Companded Data: Hybrid Representations

A wider numerical representation (larger number of bits) provides a greater dynamic range than a narrower one. A hybrid representation can provide high dynamic range in smaller number of bits, at a cost of increased quantization noise.

Early experiments in the telecommunications field determined that the dynamic range provided by 12 or 13 bits is enough to accommodate the range of signal amplitudes necessary for intelligible voice communications. However, the quantization noise provided by as few as 4 bits provides acceptable audio quality.

These results led to the development of data encoding methods that compress 12 or 13 bits down to 8 bit values, and decoders that expand 8 bits back to 13 bit values. A component or algorithm that performs both data compression and expansion is called a *compander*.

Companded data is similar to floating point data. Within an 8 bit sample, one bit is the sign bit; three bits are the "interval size" (analogous to the exponent within a floating point value), and the remaining four bits are the "interval number" (analogous to the mantissa). Companding therefore saves transmission bandwidth (8 bytes per sample instead of 12) by providing higher dynamic range (3 bits of interval size) but increasing quantization errors (since only 16 amplitude levels x 8 intervals can be represented).

U.S. telephone systems use μ–law companding, while European systems use A-Law companding. Both result in 8 bit values; the interval sizes and boundaries between intervals differ slightly. These companding methods are described in reference G.711[1].

Before operating on μ–law or A-law data with any algorithm other than data movement, the data must first be converted back to linear data.

1.7 Choosing a Representation

For counters, addresses, scaling factors, or magnitudes: unsigned integers are a natural choice.

For values where the range of the number can be well-defined, fixed-point numbers are recommended. The number of bits and location of the radix point must be determined by analysis of the algorithm. Whether the fixed point values are signed or unsigned must also be determined by analysis of the algorithm.

Signed values should always be 2's complement representation, with rare exceptions.

Within processors that support fast floating point, single precision values provide 23 bits of precision, and 128 bits of dynamic range: this greatly exceeds the precision and dynamic range provided by most sensor devices. On rare occasions, double precision may be necessary when an algorithm requires stability in order to converge. [3]

Within FPGAs, floating point should be regarded as a last resort, due to implementation complexity.

Amches Pointer

Each additional bit of width provides 6 dB more dynamic range, and also a 6 dB improvement in quantization noise. Audio CD players have dynamic range of 96 dB and SNR of 96 dB, suggesting that the data samples are 16 bits wide.

---

[3] Eigenvalue problems, root finding, matrix inversion are examples of functions that may require double precision.

	Largest positive value	Most negative value	How to negate	Notes
Unsigned	$2^b - 1$ (e.g., 255 for b=8)	0	n/a	
Sign Magnitude	$2^{b-1} - 1$ (e.g., 127 for b=8)	$-(2^{b-1} - 1)$ (e.g., -127 for b=8)	Invert sign bit	Can have -0
Two's complement	$2^{b-1} - 1$ (e.g., 127 for b=8)	-2^{b-1} (e.g., -128 for b=8)	Invert all bits, then add 1	The most negative value will overflow when negated

Amches Pointer

When an FPGA provides data to a processor, the data should be provided in a standard-width field (16 or 32 bits), left justified, so that the sign and MSB are in the most significant bit positions. Otherwise, if the data from the FPGA is right-justified, the first thing the software will have to do is sign extend the data, which is a waste of processor cycles.

Exercises

1. A system requires a signal to noise ratio of at least 48 dB. What is the minimum number of bits required to represent a sample?

2. An algorithm computes the sum of 12 separate 8-bit, two's complement samples. In order to represent the sum of all samples with no loss of data or overflow, the sum must be at least what width, in bits?

3. Re-answer the above question when the values are unsigned 8-bit values.

4. Suppose we generate an adder for two 8-bit values without any concern for overflow. What does the result look like if overflow occurs?

5. Describe the steps required to add two floating-point numbers. Compare this with the steps required to add two fixed-point numbers. Estimate the complexity between the two addition methods.

References

[1] International Telecommunication Union, *ITU-T Recommendation G.711*.

The "How to" Basics

2

In this chapter...

Having chosen a data representation, the designer must be aware of the nuances of operating on the data. For example, when adding two unsigned values, what does the result look like if an overflow occurs? Is detecting overflow important? What happens to the result when no logic is in place to detect an overflow?

For floating point models of a system, a designer almost never need worry about overflow, loss of dynamic range, loss of precision, or saturation. The FPGA developer, however, must consider all of these matters when designing a system.

A Brief History of Overflow: Mainframe computers developed before the late 1970s generated an interrupt when overflow occurred. Processors often had an overflow status indicator on a front panel. Some languages from that era provided means to detect and handle the exceptions (e.g., the "size error" clause in COBOL). Alternately, programs were allowed to abort in an error state when overflow occurred.

This trend changed in the late 1970s and early 1980s: the detection and handling of overflow became "optional".

One explanation for this transition is the increased data width provided by processors: processors supporting 32 bit data operations made overflow less likely than with 16 bit processors. Processors like the Motorola 680x0 series and Intel 386 series still detected overflow and provided an exception, but it can be disabled, and it was rarely used.

Other reasons why overflow detection became less important may be these:

- *Overflow may indicate a* data entry error *– an operator keys in 0 for a value that should never be 0, for example. In this case "handling" the error in software after overflow is not as efficient as detecting the data error during entry, and requesting re-entry. When data entry was by punched cards, such an error would not be detected until run-time. The payroll application stops; the operator figures out which input card caused the problem, and replaces the card with a corrected card. When data entry became real-time, checking and requesting the operator re-enter the data became more practical. Thus, input data can now be better filtered to eliminate values that might subsequently cause overflows.*

- *Overflow may indicate an* algorithm or implementation error. *For example, a payroll system written in the late 60s may have never anticipated that an individual's annual gross income would ever exceed $100K. In this case "handling" the error requires re-designing the algorithm to use wider data types or correcting computational steps. An exception only serves to notify the operator that the algorithm is broken.*

- *Alternately, "handling" the error may require replacing the overflow result with a saturation value. This became an automatic feature for many processors, especially with floating point: the product of two huge numbers is replaced by the IEEE value representing infinity.*

2.1 Overflow

Having grown up on languages or tools (such as spreadsheets) that accommodate values in excess of 10^{300}, few users of software have seen an overflow; fewer have had to detect or handle an overflow condition.

Overflow occurs when the result of an operation is too large to be contained within the numerical representation of the result. For fixed-point and integer operations, overflow can occur frequently.

Unless the user of a result is "immune" to overflow, a design that does not explicitly handle overflow invites disaster. Generally, overflow is "handled" by one of these methods:

- *Design it out:* for example, when adding two values, provide an extra bit of width in the result. Overflow cannot happen when an adder is constructed in this fashion.

- *Saturation:* if overflow is detected when the result would have been positive, replace the result with the largest positive representable value. Likewise, if the result was supposed to be negative and an overflow occurred, replace the result with the most negative value that can be represented. Although the saturation value is the "wrong" result, it is a far better result than that which was otherwise generated by the operation.

- *Ignore Overflow:* On rare occasions, such as when incrementing an index to a circular buffer, overflow need not be detected or handled.

Operations

This section describes how to perform common DSP arithmetic functions using fixed-point arithmetic.

Two's complement representation is assumed unless otherwise noted.

2.2 Addition

1. Align the radix points:

```
A = signed 3.2 bits =  →101 .10
B = signed 4.3 bits =   0101.101
```

2. Zero pad the operand having fewer bits on the *right* of the radix point, until both operands have the same number of fraction bits

```
A = signed 3.2 bits =   →101 .10
B = signed 4.3 bits =   0101.101

A' = signed 3.3 bits =  →101.100
B  = signed 4.3 bits =  0101.101
```

3. *Sign extend* the operand having fewer bits on the left of the radix point, until both operands have the same number of bits left of the radix point:

```
A" = signed 4.3 bits =  1101.100
B  = signed 4.3 bits =  0101.101
```

4. To *prevent* overflow, sign extend both operands by one more bit. This step is optional; a designer may instead choose to *detect* overflow (see step 6) and handle it after addition.

```
A"' = signed 5.3 bits = 11101.100
B'  = signed 5.3 bits = 00101.101
```

5. Add

6. Unless overflow was prevented (step 4), check for signed or unsigned overflow. If overflow occurs, force the result to a saturation value.

> Unsigned addition overflow: check for carry out. If carry out is set, force result to all 1s.

> Signed addition overflow: if signs of A and B are the same, and sign of result differs, then overflow has occurred. If overflow and A and B were negative, set result to 1000... . If overflow and A and B were positive, set result to 01111....

2.2.1 Signed Addition Summary

When designing an adder, consider the following:

- Do both operands have the same number of bits to the left of the radix point? If not, sign extend the operand having fewer integer bits until it has the same number of integer bits as the other operand.

- Is bit growth acceptable? If so, sign extend both operands by 1 bit.

- If bit growth is not acceptable, is saturation required? If so, perform signed overflow detection and select a saturation value based on the sign of the expected result.

- Is truncation of digits on the right required or desirable? If so, perform round to nearest processing on the result. Remember that rounding can result in overflow.

2.2.2 Unsigned Addition Summary

When designing an adder for unsigned operands, consider the following:

- Do both operands have the same number of bits to the left of the radix point? If not, zero pad the operand having fewer integer bits until it has the same number of integer bits as the other operand.

- Is bit growth acceptable? If so, zero pad both operands by 1 bit on the left.

- If bit growth is not acceptable, is saturation required? If so, perform unsigned overflow detection (carry out of the adder) and select a saturation value (all 1s) if overflow is detected.

- Is truncation of digits on the right required or desirable? If so, perform round to nearest processing on the result. Remember that rounding can result in overflow.

2.3 Negation

To negate a sign-magnitude number, simply invert the sign bit.

To negate a two's complement number, perform the following steps:

- Invert all of the bits in the number

- Add 1

Why this works:

> Suppose we need to subtract the value 5 from an arbitrary 8 bit number. It is not clear how to do this. But, it is clear how to subtract 5 from the value b'11111111':

```
  b'11111111'
  b'00000101'
  b'11111010'
```

> Therefore, inverting all of the bits of a value x is equivalent to subtracting x from the largest possible value that can be represented. This requires no carries, and no complicated logic.

Computing the two's complement requires that we then add 1:

```
    b'11111111'
    b'00000101'
  + b'00000001'
    b'11111011'      = -5
```

If we re-order the operations, it looks like this:

```
    b'11111111'
  + b'00000001'
  - b'00000101'
    b'11111011'
```

If we then combine the first two values, we get:

```
    b'100000000'
  - b'00000101'
    b'11111011'
```

Note that we needed an extra bit to compute b'11111111' + 1. Our representation for -5 within an 8 bit field is not really -5; rather it is 256-5, which is easy to compute.

When we add -5 to another value, we are adding 256-5, and ignoring the resulting overflow:

$23 - 5 = 23 + 256 - 5 = 256 + 18$, but we ignore the 256, giving 18.

This is known as *modular arithmetic*, since the result of adds or subtracts is always modulo (that is, remainder) 2^n. For 8 bit values, $(23 + 256 - 5)$ modulo 256 is 18.

Amches Pointer

Negation by complement is not unique to binary. To negate a 4 digit number in base 10, subtract 9999 minus the number, and add 1. When adding this result to another number, ignore the ten thousands (i.e., overflow) digit.

Example: to subtract 57,

$1234 - 57 = 1117$;

The 4-digit base 10 complement of 57 is $(9999-57)+1 = 9943$.

$1234 + 9943 = 11177$; ignore the overflow digit to give 1117.

2.4. Subtraction

Steps for subtraction are the same as for addition, except:

- Before adding, invert all bits of operand B, and set the carry in to the adder to '1'.

2.4.1 Signed Subtraction

When designing a subtractor, consider the following:

- Do both operands have the same number of bits to the left of the radix point? If not, sign extend the operand having fewer integer bits until it has the same number of integer bits as the other operand.

- Is bit growth acceptable? If so, sign extend both operands by 1 bit.

- If bit growth is not acceptable, is saturation required? If so, perform signed overflow detection (inverting the sign bit of the 2nd operand) and select a saturation value based on the sign of the expected result.

- Is truncation of digits on the right required or desirable? If so, perform round to nearest processing on the result. Remember that rounding can result in overflow.

- Perform subtraction by inverting the 2nd operand (including all sign extension bits), and setting carry in on the adder to 1.

2.4.2 Unsigned Subtraction

When designing a subtractor for unsigned operands, consider the following:

- Determine if the result must be signed or unsigned. If unsigned, determine if a negative result from operation should be forced to zero.

- Do both operands have the same number of bits to the left of the radix point? If not, zero pad the operand having fewer integer bits until it has the same number of integer bits as the other operand.

- Is bit growth acceptable? If so, zero pad both operands by 1 bit on the left.

- If bit growth is not acceptable, is saturation required? If so, perform unsigned overflow detection (carry out of the adder) and select a saturation value (all 1s) if overflow is detected.

- Is truncation of digits on the right required or desirable? If so, perform round to nearest processing on the result. Remember that rounding can result in overflow.

- Perform subtraction by inverting all bits of the 2nd operand and setting the carry into the adder to 1.

2.5 Comparisons

2.5.1 Unordered Comparisons

An unordered comparison determines if two values are equal or not equal. For signed or unsigned values, this can be done by exclusive-OR of corresponding bits, inverting the result of the exclusive OR, and ANDing the results together. The resulting bit value will be 1 if the values are equal. Overflow cannot occur. Shorter values must be sign extended or zero padded as needed.

2.5.2 Ordered Comparisons

An ordered comparison determines which of two values is higher than the other. Ordered comparison is performed by subtraction, and examining the result. The comparison (subtract) can be signed or unsigned.

For signed ordered comparison, use signed subtraction, check for overflow, and examine sign bit of the result, and determine whether or not the result is zero.

For unsigned ordered comparison, use unsigned subtraction. Though the inputs are unsigned, the result will be signed. Therefore, use zero padding on the input operands, but check for overflow using signed testing. Both input operands are considered positive.

2.6 Overflow Testing

Overflow can occur during addition, subtraction, comparison, negation, left-shifting, rounding, and truncation of leading bits. In most cases, overflow should be detected and the result replaced with a saturation value. An exception is circular buffer indexing, where overflow can be allowed to "wrap" modulo 2^n.

For signed addition, overflow can only occur when both inputs have the same sign. The result is expected to have the same sign as both inputs. When both inputs have the same sign, and the result sign differs, overflow has occurred. The result should then be forced to a saturation value.

For unsigned addition, overflow is indicated by a carry out of the adder.

For subtraction, the rules for overflow detection are the same as for addition, bearing in mind the fact that the second operand (and therefore its sign) will be negated before the addition occurs.

Therefore, signed subtraction overflows when the signs of the two input operands differ, and the sign of the result is not equal to the sign of the first (i.e., non-negated) operand.

Overflow occurs during negation in two's complement representation only when the input is the most-negative value: b'1000...000' cannot be negated.

Because left-shifting by N is equivalent to multiplication by 2^N, overflow can occur during left shifting. Overflow occurs during left-shifting signed values when the sign bit changes during the shift: shifting b'11011000' two bits to the left results in a positive value: b'01100000'. Also, shifting b'10100011' two bits left, though it results in a negative value b'10001100', results in overflow, since the sign changed from 1 to 0 to 1 during the shift.

Overflow occurs during truncation when left-most bits that are truncated do not *ALL* have the same value as the resulting sign bit:

b'11010011' → b'1010011'   OK, no change in sign

b'11010011' → b'010011'   overflow, sign changed

b'11010011' → b'10011'   overflow, truncated a '110'

2.7 Multiplication

Given two binary values, one having M integer bits and N fraction bits, and the other having P integer bits and Q fraction bits, the result will require $(M + N + P + Q - 1)$ bits.

The fraction portion of the result will have $N + Q$ bits. The integer portion of the result will have $M + P - 1$ bits. For unsigned values, the integer portion of the result will require $M + P$ bits.

Generally, FPGA multipliers are constructed to produce the full-width of the result determined by the size of the input operands. When this is the case, overflow can occur only in the case where the multiply is signed, and both inputs are the largest negative value.

In contrast to adders, where there is no difference between a "signed" adder and an "unsigned" adder, there is a difference between a signed multiplier and an unsigned multiplier.

The number of bits in the product can be reduced by applying round to nearest (to reduce bits to the right of the radix point), or truncation with overflow detection and saturation (to reduce bits to the left of the radix point), or both.

Example: The product of the unsigned integers $6 \times 9 = 110 \times 1001 = 3$ bits $\times 4$ bits $= 110110 = 6$ bits $= 54$ decimal.

FPGA multipliers often generate a result that accommodates the full width of the result; overflow cannot occur. However, the next stage of processing within an FPGA design may accept a "narrower" (i.e., fewer bits, not as wide) value. The designer must therefore choose: truncate high bits? Discard low bits?

Before discarding high bits, overflow detection should be employed. Before discarding low bits, rounding is often desirable; this can also result in overflow.

2.7.1 Signed Fractional Multiplication

Since fractional representation is a preferred mode of operation, it is helpful to take a look at some of the nuances of signed fractional multiplication.

An FPGA multiplier block will often provide a signed product from two signed inputs assuming that the inputs are signed integers. The output will have a width equal to the sum of the widths of the two input parameters.

Radix Point Location for Signed Fractional Multiplication

A second consideration is the location of the radix point within the result. If a signed fractional multiply is implemented using an 8 bit by 8 bit multiplier that gives a 16 bit signed result, the resulting fraction must be chosen from the proper place within the output field. For signed multiplication, that location is the second most significant bit, and the bits that follow:

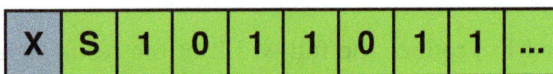

X	S	1	0	1	1	0	1	1	...

Green shaded cells are the result.

The multiplier will treat the two input values as signed integers. If the multiplier provides an output width equal to the sum of the input widths, then the fractional result will be in the positions indicated above. The high bit is a "don't care" unless the inputs were both -1.0 and -1.0, in which case the high bit, in conjunction with the next bit, indicate overflow.

The XS bits indicate the following results:

XS bits	Meaning
00	Result is $>=$ 0
01	Overflow: both inputs were -1.0
10	Cannot occur
11	Result is < 0

Overflow Detection for Fractional Multiplication

Overflow occurs for signed fractional multiply only when both inputs are -1.0. In this case, the product, 1.0, cannot be represented as a signed fractional value. This is readily detected by looking at the XS bits and saturating the result when XS = 01.

2.8 Division by a Constant

Division by a constant k is easily accomplished by converting the constant to a multiplier, and then scaling the result.

Suppose we wish to divide by the constant k. Rather than dividing by k, we can multiply by $1/k$. However, $1/k$ is not an integer value.

We can approximate an integer multiplier to $1/k$ by scaling it up: multiplying it by 2^N: the product must then be divided by 2^N in order to compensate. Since division by 2^N is shifting right N bits, we can replace division by a constant with multiplication by a new constant, followed by a right shift:

$$x/k = x * (2^N/k) / 2^N$$

The multiplier is $(2^N/k)$, where 2^N is large enough to provide an integer value having a reasonable precision. This constant will likely have fractional bits; rounding to nearest is recommended.

Example: to divide by 12, multiply by $(65536/12) = 5461$, and then shift the resulting value right by 16 bits. Since $(65536/12)$ is actually 5461.3333, the result will be off by a factor of $0.3333/5461$, or 0.00006 = 0.006 percent.

If the precision needs to be improved, use a larger 2^N value: to divide by 12, multiply by $(131072/12) = 10923$, and shift by 17 bits. The result will be off by about 0.003 percent.

2.9 Right Shifting

Right shifting n bits is equivalent to division by 2^n.

It is critical to remember that when right shifting a two's complement value, the sign bit must be shifted into the high order bits, not a zero. Otherwise, when right shifting a negative value, the result will *not* be equivalent to dividing by 2^n.

This type of shifting is also referred to as "algebraic shifting" or "arithmetic shifting." Right shifting with zero fill is also called "logical shifting," and is appropriate when shifting an unsigned value.

Example: 11010110 = (decimal -42) shifted right 3 bits is 11111010 (-6). Here we sign fill the value, since the numbers are two's complement and the original value is negative. For negative values, the result will always be rounded down (i.e., rounded towards – infinity), towards the next negative value.

Example: 00101010 = decimal 42. Shifting this right 3 bits gives 00000101 = 5.

2.10 Left Shifting

Left shifting n bits is equivalent to multiplication by 2^n.

When shifting two's complement numbers left, if the sign bit changes, an overflow has occurred.

When shifting unsigned values, shifting a 1 out of the high-order bit position signals an overflow.

Example: 11010110 = (decimal -42) shifted left 1 bit is 10101100 (-84). The input and output values have the same sign, so overflow has not occurred.

Example: 00101010 = decimal 42. Shifting this left 1 bit gives 01010100 = 84.

Amches Pointer

When left-shifting a value, be sure to employ overflow detection and have an overflow strategy.

2.11 Low-Order Truncation and Rounding

Truncation of low bits is often required in order to control bit growth. If low-order bits are nonzero before truncation or rounding, the result will add some "noise" to the result.

Truncation is easiest to implement: simply discard low bits. Truncation of a signed, positive value moves the value closer to 0. Note, however that truncation of a negative, 2's complement value makes the value more negative, moving it farther from zero. For example, consider a 4.2 bit values representing 2.25 and -2.25:

x	*4.2 bit binary*	*Truncated binary*	*decimal*
2.25	0010.01	0010	2
2.50	0010.10	0010	2
2.75	0010.11	0010	2
-2.25	1101.11	1101	-3
-2.50	1101.10	1101	-3
-2.75	1101.01	1101	-3

Mathematically, truncating a two's complement value should be thought of as subtracting low-order bits, forcing them to zero, rather than simply discarding them. It is equivalent to the C *floor* function. The result will have a negative bias of -1/2 times the value of the least significant bit remaining after truncation.

Rounding provides slightly more accuracy, and does not introduce the slight negative bias introduced by truncation. While there are multiple ways to round, *rounded to nearest*[4] is the most common.

---

[4] Other forms include round towards +infinity, round towards –infinity, round towards zero. Truncation is equivalent to round towards –infinity.

For positive or unsigned values, rounding is accomplished by adding ½ to the value and then discarding low bits. Because adding ½ may result in overflow, saturation or additional bits on the left must be provided.

When rounding a negative decimal value to nearest, we think of subtraction one-half and then dropping the decimal portion, so that -2.75 – 0.5 becomes -3.25, which rounds to -3.0. However, when rounding two's complement values, we can take advantage of the behavior provided by truncation, and add ½ and then truncate:

x	4.2 bit binary	x + 0000.10	Truncated binary	decimal
2.25	0010.01	0010.11	0010	2
2.50	0010.10	0011.00	0011	3
2.75	0010.11	0011.01	0011	3
-2.25	1101.11	1110.01	1110	-2
-2.50	1101.10	1110.00	1110	-2
-2.75	1101.01	1101.11	1101	-3

Amches Pointer

To round the least significant N bits off of a signed, 2's complement value, add 2^{N-1} with saturation detection and shift the result right N bits. Overflow can only occur when N is a large, positive value.

Examples:

These examples use fixed point values having four bits to the right of the radix point. Rounding is to the nearest integer:

```
3.75 =   0011.1100          input value
         0000.1000          rounding value
         0100        =      4.0

-3.75 =  1100.0100          input value
         0000.0000          rounding value
         1100        =      -4.0

-3.25 =  1100.1100          input value
         0000.1000          rounding value
         1101 =             -3.0
```

Rounding to nearest will result in overflow if the input value is a large positive value. Overflow detection with saturation should be provided when rounding to nearest.

Round to Even: many processors support a mode called *round to even*. When a value to be rounded is exactly halfway between two integer values, this rounding method chooses the even integer value. When the probability of even versus odd values are equal, this method eliminates a slight bias. This requires more FPGA logic:

- Test the bits to be truncated to see if they match b'100..0', which is one half. If not, perform rounding as described above.

- Otherwise, if the bits to the left indicate an odd value, add 1 to the bits to the left of the bits to be discarded, with saturation, and use this as the result.

- If the bits to the left indicate an even value, choose the unaltered bits to the left as the result.

For two's complement representation, and odd value is indicated by a 0 sign bit and 1 least significant bit, or 1 sign bit and a 0 least significant bit.

2.12 High-Order Truncation

Truncation of high bits is sometimes required in order to control bit growth. For example, when multiplying two 8-bit integers, the result will require 15 bits for representation. However, if the inputs to the multiplier are known to be limited so that the result is less than 4096, high order bits may be discarded without consequence.

If overflow detection is important, the design must first check the high-order bits that will be truncated:

When truncating an unsigned result, the discarded high bits must be all zeroes; otherwise an overflow occurs.

When truncating a two's complement value, the high bits must either be all zeroes, or all 1s. Furthermore, the most significant bit that remains in the result must match the high order bits. If these conditions are not met, truncation will result in an overflow.

2.13 Estimating Magnitude

In DSP applications it is sometimes necessary to compute sqrt($x^2 + y^2$), where x and y are the real and imaginary parts of a complex value.

This algorithm computes an approximate result that is sufficient for many applications:

1. Compute the absolute value of x and y.

2. Re-order the operands so that x is the largest value, y is the smallest

3. Compute result = $\alpha x + \beta y$, where α = 15/16, and β = 15/32.

Division by 16 and 32 are simple shifts right. Multiplication by 15 can be simplified to computing 16x – x, where multiplication by 16 is a simple shift left.

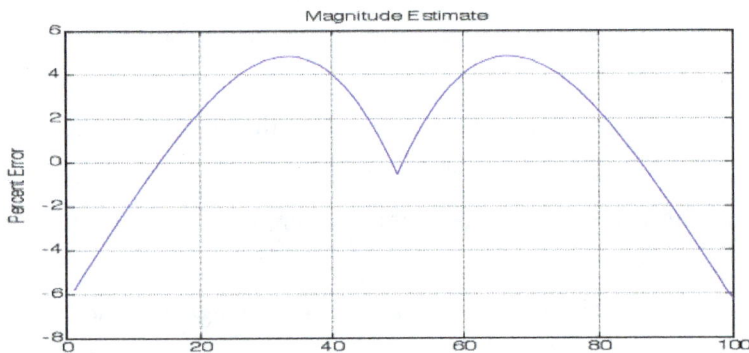

The figure to the left shows % error for x values of (1..100) and y values of (100..1), using the α and β values above. Other values for α, β, are sometimes used; see Lyons[1] for further discussion.

Figure 1. Percent Error for magnitude estimation function.

Example: The complex number (23 – i77) has a magnitude of sqrt(529 + 5929) = sqrt (6458) = 80.3616. The estimated magnitude is (77*15/16) + (23*15/32) = (72.1875 + 10.78125) = 82.96875, off by about 3.2%.

Amches Pointer

This method is very fast, but it is an estimate that will be off by a small amount. For example, when x = 1000 and y = 500, the result is 1172, versus an expected value of 1118, for a relative error of 4.8%.

The method is often good enough for determining if a signal level is greater than a threshold value; the threshold can be reduced by 6% to compensate for the curve above.

2.14 Phase Change Estimation

The phase angle of a complex value z is computed from the atan2 (y, x) function, where y = imag(z) and x = real(z).[5] For a complex sinusoid, the difference in phase angles between two consecutive samples is proportional to the frequency. For example, if the sinusoid repeats every 8 samples, then the phase angle between consecutive samples is always $2\pi/8$.

The phase difference between two consecutive samples $d\phi$ is atan2 (s[n]) – atan2(s[n-1]), where the samples s are complex. However, the arc tangent function is difficult to compute.

For phase angle deltas between $-\pi/4$ and $\pi/4$, when the samples s are normalized so that they lie on the unit circle, an estimate is given by:

$$d\phi = imag(s_n) \cdot real(s_{n-1}) - real(s_n) \cdot imag(s_{n-1})$$

Normalization can be approximated by dividing by the estimated magnitude. The plots below show the estimates and errors over a range of $-\pi/2$ and $\pi/2$. When the phase angle is smaller ($-\pi/4$ to $\pi/4$), the worst case error is 10%.

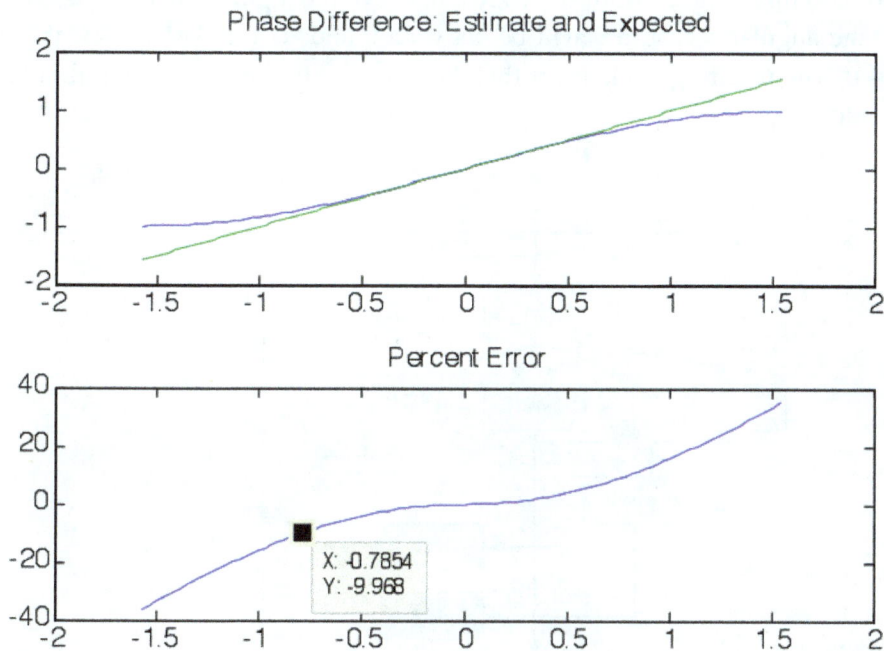

Figure 2. Delta Phase Estimation Function

---

[5] Some atan2 functions take the y value first, others take the x value first. Double check function documentation before using atan2 functions.

2.15 Rectangular to Polar Conversion

Many demodulators require conversion from rectangular to polar coordinates. Conversion involves computing the magnitude of a complex value, and also the phase, or arc tangent, of the y/x value.

While the magnitude can be estimated within a few percent, more accurate results are often desired. Also, the arctangent is difficult to compute, especially when y, x, or both are zero or extreme values.

The approach presented here requires a single divide, a single multiply, and two lookup tables. The input values x, y are signed integers; signed fractions can be used if the lookup values are adjusted. The precision of the result is approximately equal to the table sizes. The table lookup argument is the most significant bits of the value $\min(|x|, |y|) / \max(|x|, |y|)$. This value is always in the range 0..1. The square root lookup function determines a weighting factor, that, when multiplied by $\max(|x|, |y|)$, gives an approximation to the magnitude of the complex value $x + iy$. This table represents the function $\sqrt{(a^2 + 1)}$ for a in the range 0 to 1.0.

The arc tangent lookup function determines the arc tangent of the input ratio; because the input is in the range 0..1, the angular measure θ will be between 0 and $\pi/4$. If $|y| > |x|$, the value is adjusted to $\pi/2 - \theta$. The resulting angle must then be placed into the proper quadrant based on the signs of the x and y inputs.

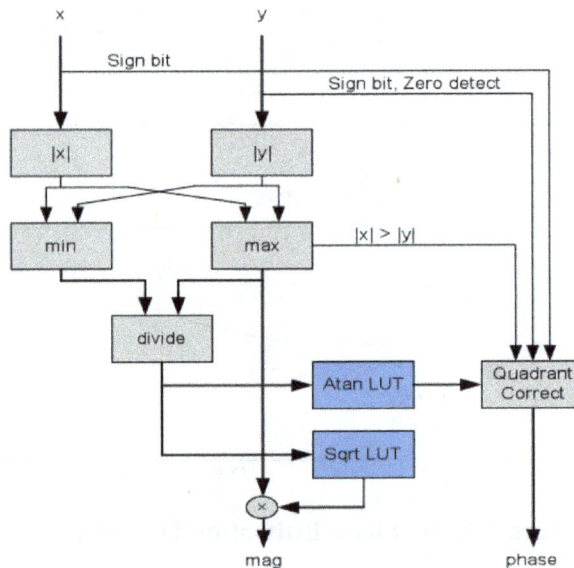

Figure 3. Polar to Rectangular Conversion.

2.16 Log Estimation

The base two logarithm of a positive integer value of length b bits can be estimated using the following method:

1. Count the leading zeroes n.

2. The integer portion of the result will be b-n-1.

3. Use the first nonzero bit, and the next k bits that follow, as an index for a table lookup to determine the fractional part of the result.

The result will be signed, fixed point, having some number of fractional bits determined by the table. Figure 4 shows the estimation of the $\log_2$ for values between 1000 and 4000 using a 16-entry table. Each table entry is a fractional value accurate to 15 bits. The table is populated with entries of $\log_2 (x)$, where x ranges from 1.0 to $1 + 15/16$ in steps of $1/16$. Table entries themselves range from 0 to 0.9221.

The absolute error over this range is as high as 0.087, occurring at $\log_2(2175)$. The relative error at this point is 0.008. For larger values of x, the absolute error will increase, but the relative error will remain bounded. A smaller relative error is achieved by using a larger lookup table.

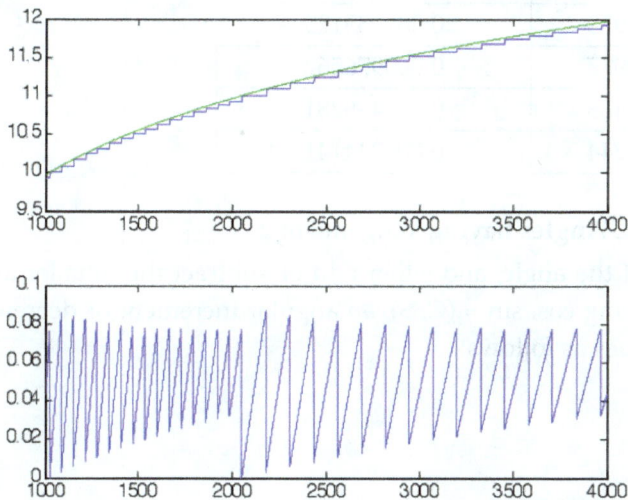

Figure 4. Log2 estimation and absolute error.

The following refinements to this method are possible:

- An input value of all zeroes should generate the most negative value representable in the result field.

- The value used for the index, as described above, will always have the high order bit set. (The exception to this is a zero input value, which is handled as a separate special case). The bit can be ignored, and the index to the table values can be based on the next k-1 bits.

- Conversion to base e is achieved by multiplying the result by log(2), approximately 0.69314.

- Conversion to an arbitrary base b is accomplished by multiplying by log(2)/log(b).

2.17 CORDIC Functions

Trigonometric and inverse trigonometric functions can be computed using CORDIC (COordinate Rotation DIgital Computer) algorithms. CORDIC algorithms particularly well suited to FPGA implementations. See Andraka [2].

CORDIC algorithms are similar to the bisection algorithm, except that rather than bisecting angles, the space being bisected is the *tangent* of the angular value. CORDIC therefore uses angles having tangent values of the form 2^{-i} for i = 1 to n. A table of arc tangents of values 2^{-i} is employed.

i	2^{-i}	*Angle (radians): $atan(2^{-i})$*
1	0.5	0.463647609
2	0.25	0.244978663
3	0.125	0.124354995
4	0.0625	0.06241881
5	0.03125	0.031239833
6	0.015625	0.015623729
7	0.007813	0.007812341
8	0.003906	0.00390623
9	0.001953	0.001953123
10	0.000977	0.000976562
11	0.000488	0.000488281
12	0.000244	0.000244141

Table 1. CORDIC Angles having Tangents of 2^{-i}.

Successive iterations halve the tangent of the angle, and either add or subtract this smaller angle to a cumulative result. For any angle having cos, sin = (C, S), an angular increment or decrement of $atan(2^{-i})$ radians changes the (C, S) values as follows:

$$C_{n+1} = W_n[C_n - (\pm S_n \cdot 2^{-i})]$$
$$S_{n+1} = W_n[S_n + (\pm C_n \cdot 2^{-i})]$$
$$W_n = 1/\sqrt{(1 + 2^{-2i})}$$

Thus, rotation by an angle whose tangent is 2^{-i} requires the addition or subtraction of a shifted value $(S_n \cdot 2^{-i})$, and then multiplication by the weighting factor W_n. Weighting factors can be pre-computed. Also, a series of rotations may be applied (for i values of 1, 2, 3.. n) without weighting, and then a final weighting factor consisting of the product of weighting factors (1, 2, … n) can be applied, reducing the number of required multiplications to 2.

28

To compute the cos, sin of 0.33 radians., use a starting angle of 0 (having cos, sin = 1, 0), and a starting radian measure of -0.33. For each successive angle in Table 1, the algorithm adds or subtracts the radian value of atan(2^{-i}) from -0.33, driving its absolute value closer to 0, while adjusting the C and S values according to the equations above.

This gives rise to the following sequence:

i	$-/+$	Angle	C_n	S_n
		0.330000000000	1	0
1	1	0.133647609001	0.89442719099992	0.44721359549996
2	-1	-0.111331054126	0.97618706018395	0.21693045781866
3	1	0.013023940421	0.94174191159484	0.33633639699816
4	-1	-0.049394869575	0.96088803114899	0.27693716181587
5	1	-0.018155036145	0.95176912722744	0.30681513726789
6	1	-0.002531307525	0.94685956431851	0.32164726869438
7	1	0.005281033536	0.94431787717697	0.32903456785571
8	-1	0.001374803404	0.94559595416057	0.32534334398472
9	-1	-0.000578319113	0.94622958559131	0.32349585986796
10	1	0.000398243077	0.94591322061919	0.32441975750228
11	-1	-0.000090038134	0.94607151592340	0.32395784719374
12	1	0.000154102486	0.94599239645934	0.32418881202335
13	-1	0.000032032174	0.94603196324045	0.32407333202136
14	-1	-0.000029002982	0.94605174134475	0.32401559020914
15	1	0.000001514596	0.94604185273312	0.32404446126617
16	-1	-0.000013744193	0.94604679714907	0.32403002577537

Table 2. CORDIC Sequence for computing sin, cos of 0.33.

The first row shows the starting values for the angle, and the cos/sin of 0 radians.

CORDIC functions require about n iterations to compute a result to an accuracy of n bits.

CORDIC functions can compute sin, cos, arctangent, arcsin, arccos, polar to rectangular, and rectangular to polar conversions.

Exercises

1. An analog to digital converter converts voltages from -1v to +1v to an 8 bit value where -1v results in a value of 0, 0v results in a value of 128, and +1v results in a value of 255. The algorithms require values of -128 for -1v, and +127 for +1v. What is the minimum amount of logic needed to convert the values?

2. Describe the differences between an adder for unsigned values, and an adder for signed values.

3. An 8 bit unsigned value is squared. How many bits wide is the result?

4. An 8 bit signed value is squared. How many bits wide is the result?

5. An FPGA functional block computes the square root of a 12 bit integer. How wide is the result if it is an integer?

6. An FPGA functional block computes the square root of a 12 bit integer. How many bits to the right of the radix point must be maintained in order to generate the original value after squaring the result?

7. When a square root functional block is provided with a negative input value, what output value should it provide?

8. Implement the following in VHDL:

    a. Functional block which rounds a 10 bit unsigned value, returning the most significant 8 bits, using round to nearest.

    b. Functional block which rounds a 10 bit signed value, returning the most significant 8 bits, using round to even.

    c. Functional block which adds two 10 bit signed values, providing a saturated, signed result.

    d. Functional block which estimates the magnitude of a complex value, where the inputs are 12 bit real and imaginary signed values. Use $\alpha = 15/16$, and $\beta = 15/32$.

References

[1] Lyons, Richard G., *Understanding Digital Signal Processing*, Second Edition, Prentice Hall, Upper Saddle River, New Jersey, 2004, pp. 479-484.

[2] Andraka, R. J., "A survey of CORDIC algorithms for FPGA based computers", ACM 0-89791-978-5/98/01, 1998.

Signals and Noise

3

A signal is a method of communicating information between two parties. Within communications systems, signal processing involves translating information into a form which can be transmitted though some medium, transmitting the signal, receiving the signal, and translating it back to information for a recipient.

Signals and noise can be analyzed. Processing a signal often adds noise, though some times of noise can be reduced by signal processing methods. The effect that processing has on a signal and noise levels is a matter of design; not accident.

Signals are not limited to radio: an infrared signal is emitted from a remote control to a TV, and ultrasonic signals are used in medical imaging. Other signals may be mechanically transmitted (vibrations). Subsonic signals within the earth are analyzed to understand seismic activity.

3.1 Periodic Signals

We are largely interested in signals that are "mostly" periodic: these signals have a repeating pattern over some period of time, such as a sine wave. Periodic signals are of interest for a few reasons:

- Often, the media lends itself to transmission of periodic signals. We cannot transmit DC voltage levels through the air, but an AC signal having sufficiently high frequency can be radiated and received. It is difficult to transmit "absolute pressure" through the air, though varying pressure, in the form of sound, is readily transmitted.

- Transmission media can often accommodate multiple, concurrent periodic signals of different frequencies, so the media can be shared. This is also true for sound: people conversing interfere with each other, but they do not interfere with bat sonar location, and visa versa.

- A rich set of signal processing tools and approaches exist for working with periodic signals. The richness of these tools is perhaps a consequence of the first two attributes above.

A signal that is strictly periodic (that is, a pure sine wave) is of limited value. In order to increase the value, systems *modulate* this signal (change it slightly) over time to represent encoded information. For a sine wave, a system may change its amplitude, its frequency, its phase, or a combination of these parameters in order to represent information. The unmodulated sine wave is called the *carrier frequency*. As the name implies, it serves to carry the information over the media.

Amches Pointer

Changing any feature of a sine wave always introduces new frequencies into the spectrum. More about this later…

3.2 Adding Signals Together

When signals are added together, the resulting spectrum is equal to the sum of the spectra of the individual signals.

This is also true with light. Adding blue and green results in cyan, but when the result is passed through a prism, the resulting spectrum is equal to the sum of the spectra of the blue and green lines.

When a signal is added to noise, the resulting spectrum is the sum of spectrum of the signal and the spectrum of the noise. Unfortunately, some noise often occupies the same portion of the spectrum as the signal we wish to process. Noise in other parts of the spectrum can be filtered out. Noise that is coincident with the signal we wish to process will "ride through" the system, with the signal.

3.3 Signals and Harmonics

A musical instrument may generate a note having a frequency of 440 hertz[6] (Chamberlin [1]). However, the note from the instrument will also have a frequency of lower amplitude at 880 Hz, and at 1320 Hz, and 1760 Hz, and so on. The lowest frequency is called the *fundamental frequency*; the other frequencies are integer multiples of the frequency, and are called *harmonics*. The number of harmonics and their relative amplitudes help give different instruments their distinct sounds. 880 Hz is the second harmonic of 440 Hz; 1320 is the 3rd harmonic, etc. Often higher harmonics have lower amplitude, though this is not always the case.

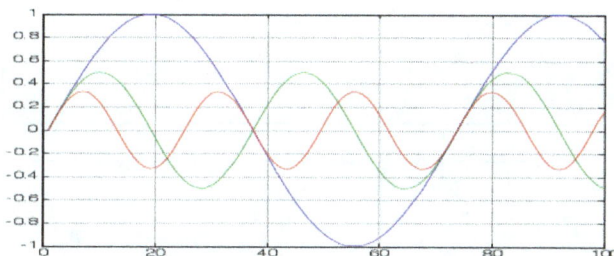

Figure 5. A fundamental and two harmonics.

When a signal includes harmonics, the signal still repeats at a rate equal to the fundamental frequency, but the shape is no longer sinusoidal. The opposite is also true: a repeating signal that is not perfectly sinusoidal contains harmonics. Figure 6 shows the sum of the signals presented in Figure 5.

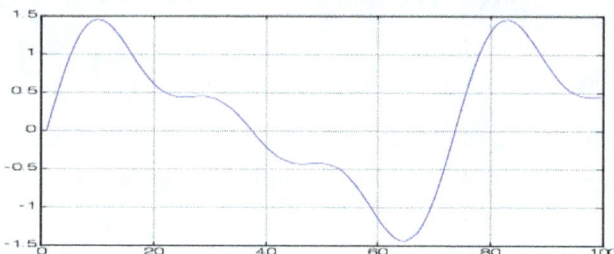

Figure 6. A signal with harmonics.

---

[6] This is middle A on the musical scale.

Addition of Signals that are not Harmonically Related

If we 440 Hz to another signal f_2 that is not a harmonic (that is, it is not a multiple of 440 Hz), the resulting waveform will differ from cycle to cycle, and the waveform shape will not repeat *exactly* until it a time equal to 1/GCF (440, f_2), where GCF is the greatest common factor of f_2 and 440.

Suppose we add 440 Hz and 350 Hz. The greatest common factor is 10; the combination of signals will repeat exactly every 1/10th of a second. After 1/10th of a second, the 440 Hz component will have completed exactly 44 cycles, and the 350 Hz component will have completed 35 exactly cycles. Both components reached their starting point at this same time, and the sequence repeats.

These signals, when added, appear to have an approximate "envelope" of about 123 Hz.

The North American dial tone is the sum of 350 and 440 Hz. Because the two signal components are not harmonically related, the signal has a "mechanical" buzzing sound, rather than the sound of a smooth, single note from an instrument.

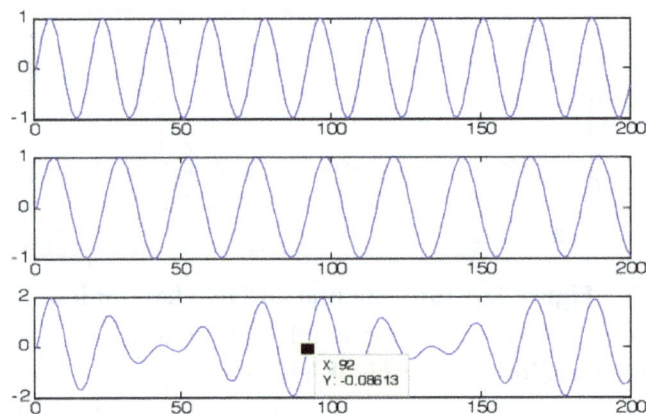

Figure 7. The sum of 350 Hz and 440 Hz.

3.4 Signals and Noise

Noise is defined as anything other than the signal that we wish to process. Noise can include other signals: in a radio receiver, we wish to tune a single signal, and eliminate the effects of other signals.

Some noise is present in the environment: an antenna will pick up both signals and noise. Some noise will be generated by the sensor. Ideally this noise is a very small fraction of the signal measured by the sensor. Some noise is generated within signal processing equipment: even resistors generate electrical noise from thermal activity.

Almost all digital and analog processing performed on a signal introduces some amount of noise. A key design objective is to keep the level of the signal well above the level of the noise, so that the effects of the noise are negligible.

External Noise	Characteristics
Astronomical	low level; VHF and up; diminishes with frequency
Atmospheric	high level, HF – VHF
Electromechanical	high level, HF and lower
Electronic	HF – UHF
Adjacent Channels	"busy" portions of the spectrum
Internal Noise	Cause
Nonlinear Distortions	Overdriving components
Thermal Noise	Physics: thermal agitation of electrons
ADC Quantization	a function of ADC precision
ADC clock jitter	a function of ADC clock source precision
Rounding, Truncation, Algorithmic Processes	Algorithm Design

Some noise may be characterized as random (atmospheric noise, thermal noise, quantization noise, rounding and truncation), while other noise is coherent (adjacent channel noise, distortions, clock jitter). Averaging and filtering may help reduce random noise; reducing other noise requires more analysis and more sophisticated methods, possibly including redesign with better components.

The *signal-to-noise-ratio* (SNR) indicates how much more powerful the signal is than the noise. This ratio is normally expressed in dB. Higher values are desirable. For audio, expect an SNR of 24 dB or so from a telephone, while a CD player and quality set of headphones provide 96 dB over a wider set of frequencies.

Amches Pointer

In sound systems, harmonics generated from nonlinear behavior (harmonic distortion) are measured separately from the signal to noise ratio. A sound system may have an SNR of anywhere from 60 dB to over 100 dB, depending on the cost of the system. It may also have a total harmonic distortion (THD) of 0.2% at certain frequencies and power levels.

3.5 Signal Processing

Signal Processing consists of applying hardware and/or algorithms to transmit, receive, detect, enhance, or store information. A radio link may incorporate the following steps:

- Add optional error detection and correction information to the signal;

- Encode or translate information (speech, music, text, video, etc.) into a form compatible with the transmission media (usually by *modulating* a *carrier*);

- Amplify and Transmit the modulated carrier signal: place the modulated carrier onto the media, often in the presence of other signals;

- At the receiver, select the signal of interest ("tune" the receiver);

- Minimize or eliminate unwanted signals;

- Amplify the signal if needed;

- Change the encoded signal back into a form that is useful to the consumer (demodulate).

- (Optional) detect and correct errors within the signal.

Similarly, an ultrasound imaging system may incorporate the following processing steps:

- Generate a precise signal source, through multiple transducers, having carefully controlled phase relationships

- Acquire, filter, and amplify returned ultrasonic signals, through multiple transducers

- Adjust for signal propagation through signaling media;

- Correlate or combine inputs from multiple transducers to form a spatial image

- Convert the resulting information to a displayable format.

3.6 Signals and Spectra

A signal may be viewed on an oscilloscope as a time-varying amplitude. However, unless the signal is a single sine wave, square wave or other simple wave, it is often impossible to recognize signal features using an oscilloscope. The combination of multiple frequencies and phases often create a "jumble" rather than a nicely repeating pattern. A more useful means to view and analyze a signal is to display its spectrum using a spectrum analyzer. This breaks the signal down into its individual frequencies, and shows the relative magnitude of each spectral component.

The spectrum of a light source shows the various colors emitted from the source. A light spectrum also indicates the amplitude of each color: a source having more blue than red will show a brighter blue region in the spectral decomposition.

For electrical signals, the spectrum is a plot of *magnitude* versus *frequency*. The spectrum will have peaks at certain frequencies; these indicate the presence of sine and cosine waves at those frequencies. It will also show a low level between the peaks. The noise floor is an approximation of the average value of the "non-peak" portions of the spectrum.

It is often desirable to show plot the spectrum in decibels (dB). The plots in Figure 8 show the dial tone spectrum plotted using linear amplitude and dB. The peaks occur at 350 and 440. Using dB allows us to see the "shape" of the noise. This mathematically generated example is a very "clean" case: most spectra have far more noise than this.

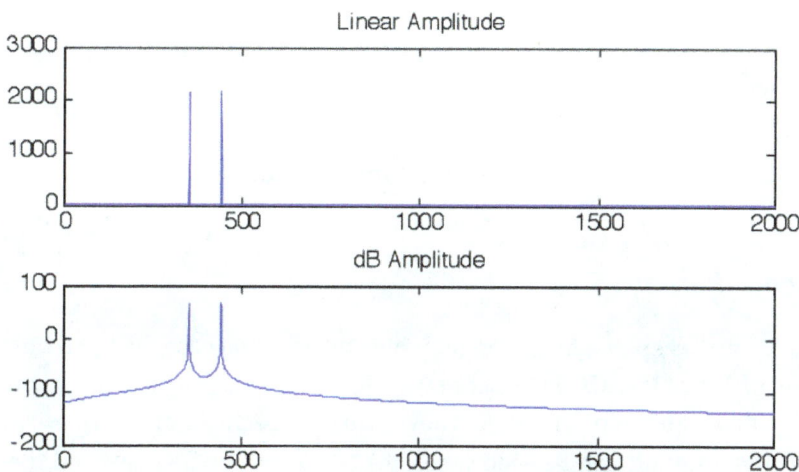

Figure 8. The Spectrum of a Dial Tone.

The figure below shows a dial tone in the presence of a very small amount of noise. This is more typical of an actual signal. The time domain plot with noise is not provided. The time domain plot for this signal plus noise would look almost exactly like the time domain plot for the signal without noise; a noise level his low is barely discernible on such a plot. The noise is also not noticeable from the spectrum with the linear scale. The dB scale, however, clearly shows the noise.

The noise floor is slightly below 0 dB; the signal levels are at 67 dB, well above the noise floor. Compare this to the noise level in Figure 8. In that figure, though it was difficult to tell where the noise floor started, it was several dB below 0.

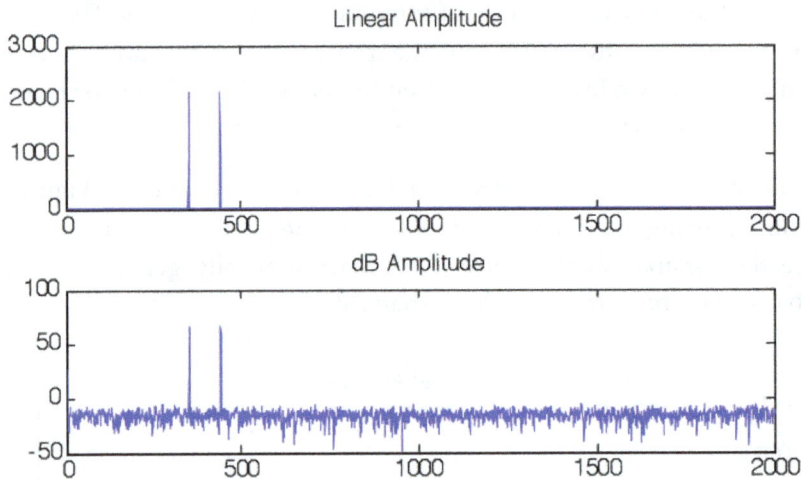

Figure 9. A dial tone with noise.

> ### 🌐 *Amches Pointer*
>
> When signal amplitudes span a large range, which is often the case, use of a linear scale would cause us to adjust the display so that we can see the largest signals; this would push the smaller ones down to the point where they cannot be seen. Alternately, adjusting the display so that small signals can be seen would cause large signals to exceed the top of the display. The dB scale allows us to view small and large signals, even though the large signal may have 10,000,000 times the power of the small signal.

3.7 Some Typical Power Levels

When using a signal generator, the amplitude may be specified in volts, or alternately decibels with respect to one milliwatt of power into a 600 ohm load (dBm). From Wikipedia[2], the following references are helpful in understanding the relative powers specified in dBm:

Power (dBm)	Power (mW)	Application
27	500	Typical cellphone transmit power
26	400	Access point for wireless networking
24	250	Max output from a 3G mobile phone (power class 3)
23	200	Max output, interior environment from WiFi 2.4 GHz antenna

20	100	Bluetooth class 1 radio, 100 m range
15	32	Typical WiFi transmitter in lap tops
4	2.5	Bluetooth class 2 radio, 10 m range
0	1.0	Bluetooth class 1 radio, 1 m range
-10	100 uW	Typical max rx power for wireless network
-70	100 pW	Typical range of wireless rx power for 802.11 network
-111	8 fW	Thermal noise floor for single channel GPS rx channel, (2 MHz)
-127.5	0.18 fW	Typical rx signal power from GPS satellite

Table 3. Typical power levels, in dBm and milliwatts.

From the table above, a spectrum analyzer display can span a huge range of signals if the Y axis units span 120 dB.

3.8 Distortion

For most signal processing algorithms and devices, we require that the spectrum of the signal that goes in look like the spectrum of the signal that comes out, with the possible absence of signals in which we have no interest. For example, if we amplify a modulated signal having a frequency of 455 KHz, we expect the output to also be 455 KHz, but with a larger amplitude.

If an amplifier distorts a signal, two results ensue: first, the output signal energy at the expected frequency is diminished below the expected value, and second, energy spikes appear at new places in the spectrum (i.e., new frequencies show up in the output; these are harmonics of the input signals). In an audio system, this is called *harmonic distortion*.

If the original signal was a single sine wave, the resulting signal has the original signal plus harmonics. If the original signal was the sum of multiple sine waves, the resulting signal has harmonics of the original components, and also harmonics of the sums and differences of all of the original components.

An amplifier may *clip* a signal (cut off the peaks) if the input is too large, or the power to the amplifier is too small, or if the amplifier is driven too hard (i.e., turned up too loud). Clipping generates rich sets of harmonics, illustrated in Figure 10. The dial tone, without noise, has been clipped when the value exceeds the range -1.7 to +1.7. Although the time domain looks approximately the same, the frequency domain now includes harmonics throughout the spectrum.

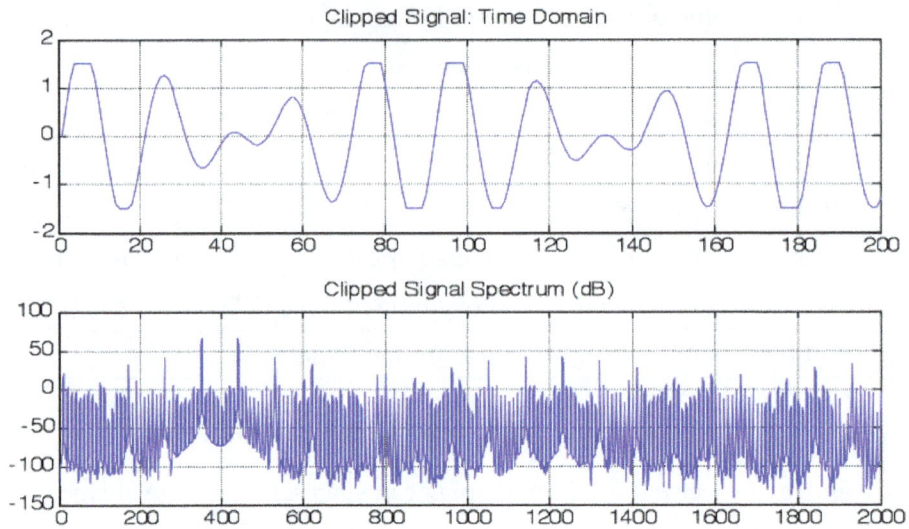

Figure 10. Effect of clipping on the spectrum of a signal.

Any nonlinear process applied to a signal distorts the signal, and produces harmonics. Other examples of nonlinear processes include rectifying a signal, multiplying a signal by a time-varying value, and passing a signal through a nonlinear amplifier.

The figure below represents an amplifier. An input voltage x is mapped to an output voltage y based on the transfer function in the x-y plane. Note that the transfer function is *almost* linear over the range of the input signal, but it has a slight curvature. When a perfect sine wave is provided as the input to this amplifier, the resulting sine wave will be slightly distorted.

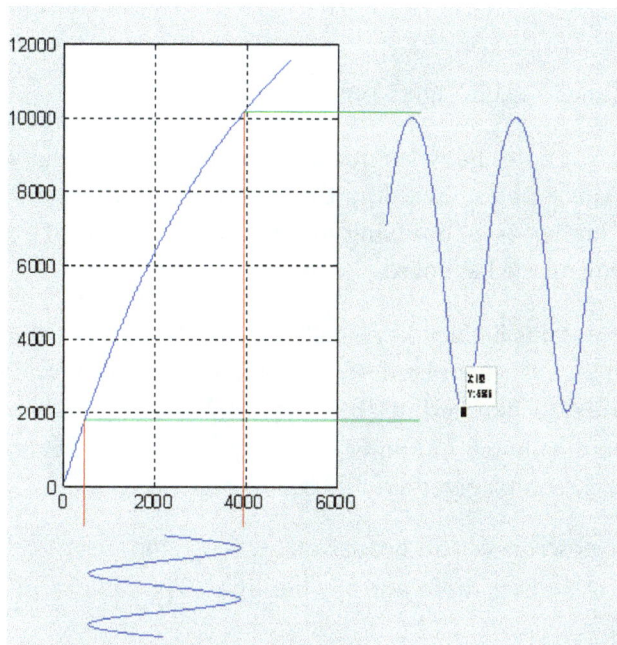

Figure 11. Distortion due to non-linear amplification.

Though the output looks sinusoidal, the resulting spectrum shows that the output has numerous harmonics having significant amplitude:

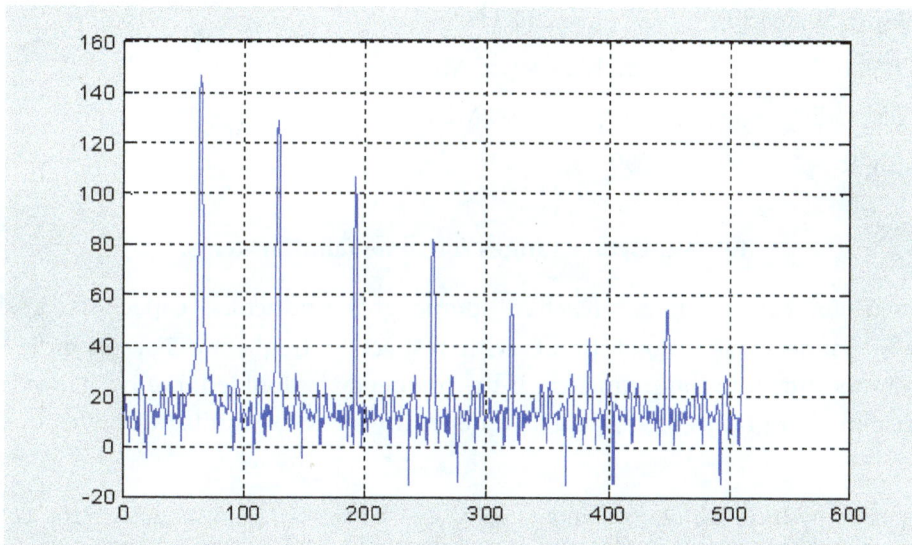

Figure 12. Harmonics resulting from nonlinear amplification.

3.9 Octaves

The term *octave* is sometimes used in signal processing.

On a keyboard, the range of notes between middle C and the next higher C is one octave, named so because this range spans 8 notes, including the C and the next higher C. Mathematically (and physically), an octave higher is a doubling in frequency; an octave lower is a halving in frequency. This is also true on the keyboard.

A keyboard octave also contains 5 black keys, called *accidentals*. From one octave to the next there are a total of 12 keys. In an even-tempered scale, the frequency from one key to the next key on the right has a ratio of $1:^{12}\sqrt{2}$ (Chamberlin[1]). A step from any key to the adjacent key (including accidentals) is also called a half-step.[7] Moving from any note on the keyboard to the note 12 keys to the right (including accidentals) gives a doubling in frequency.

Octaves are not only descriptive of acoustical waves: the term can also be applied to electromagnetic waves, including radio and visible light. The table below may help the reader appreciate octaves.

Object	Range	Octaves
Piano	88 keys	7+
Guitar	2 octaves from string 1 to string 6, plus ~20 half-steps	3+
Human Hearing	20 Hz – 20 KHz[a]	10
AM radio	550 KHz – 1.65 MHz	1 ½
FM radio	88 MHz – 108 MHz	< 1
Light	390 – 750 nM[b]	< 1

Table 4. Octave ranges for some familiar items.

Octaves are useful because many reactive components (inductors, capacitors, springs, etc.) respond in a manner that is logarithmic with respect to frequency. For example, inductive reactance is proportional to the frequency: if the frequency is doubled, the reactance is doubled. This gives rise to filter graphs having y-axis units of dB and x-axis units of octaves. A circuit that

---

[7] Confused about why there are 12 notes in an octave, and a half step is a ratio of 1 to the 12th root of 2? Much of this is probably due to the fact that music was invented before the spectrum analyzer.

[a] The upper limit for an adult is lower, probably due to the size and mass of the auditory components in the ear.

[b] These are wavelengths; the frequency is inversely proportional to the wavelength.

generates half the output voltage when the input voltage remains the same but the frequency doubles has an attenuation of 6 dB/octave.

3.10 Bandwidth

A signal occurs at a specific place within the spectrum, and requires a specific width based on the information it carries and the modulation method employed. The bandwidth is a function of the modulation type and the modulating signal. The specific place is called the center frequency, and the width is called the bandwidth. The center frequency for an FM broadcast station, for example, is 98.7 MHz. The allocated bandwidth is about 150 KHz, and stations are spaced 200 KHz apart, so that the next station might operate at 98.9 MHz.

Filters also have a specific center frequency and bandwidth. If the center frequency and bandwidth of a filter match the center frequency and bandwidth of a signal, then the filter will be helpful in that it will suppress un-interesting signals. If the center frequencies match but the bandwidth of the filter is larger than that required by the signal, then the filter will admit more noise (possibly including other signals) than desired.

If the filter bandwidth is narrower than that of the signal, the system may not capture enough information to faithfully demodulate the signal. If the center frequency of the filter is not aligned to that of the signal, then the filter will compromise the information contained within the signal: this is the case when a radio is slightly "off-tuned" from the desired station.

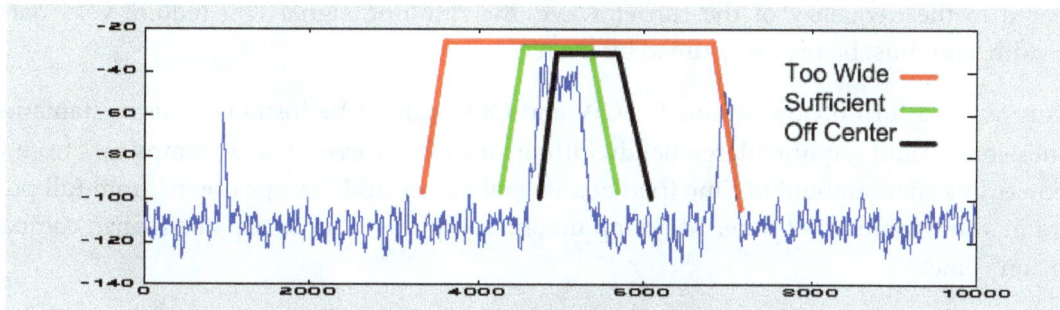

Figure 13. Signal and Filter Bandwidth matching.

3.11 Introduction to Modulation

Radio superimposes an audio information signal (or video or data) onto an RF carrier wave by changing the carrier wave in a manner that represents the information.

There are multiple modulation methods; see Shanmugam[3]. These methods reflect tradeoffs between signal quality, power requirements, cost of transmission electronics, and cost of receiving electronics. Also, the feasibility of more complex methods has changed with the evolution of electronics engineering.

An information signal itself requires some amount of signal bandwidth. The modulated carrier may require a multiple of the information signal bandwidth. Signals having wider bandwidth require more transmission power to provide an acceptable SNR at a receiver.

A brief summary of some analog modulation methods, and a digital modulation method, follow.

CW and OOK Modulation

Continuous Wave (CW) modulation, also known as on-off keyed (OOK) modulation, encodes information by turning a carrier on and off. For Morse code, where the keying rate is low compared to the frequency of the carrier wave, the resulting signal will require very narrow bandwidth, and must be precisely tuned by the receiver.

The turn on and turn off transitions for CW and OOK cannot be instantaneous; instantaneous transmissions would require a large bandwidth. Instead, the carrier wave "ramps up" from 0 to full power in a short amount of time (perhaps several mSec), and "ramps down" from full power to zero in a short amount of time. The ramp up and ramp down times are small when compared to the "on" time.

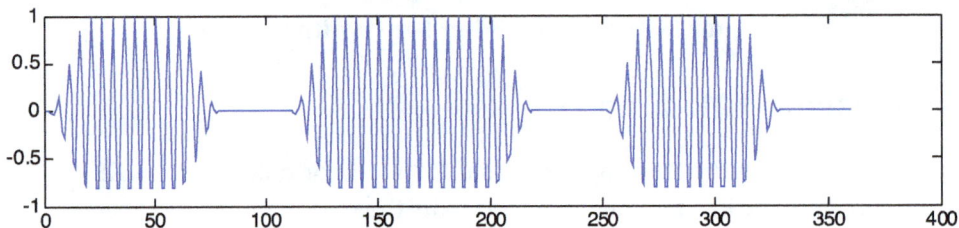

Figure 14. A CW signal; note the ramp up and ramp down transitions.

Amplitude Modulation

One of the simplest forms of modulation capable of carrying voice and music is amplitude modulation (AM). In AM, the amplitude of the carrier wave is varied in proportion to the amplitude of the information signal. An underlying assumption is that the carrier wave has a much higher frequency than that of the information signal.

An AM modulator is essentially an amplifier for the carrier signal, the "volume" of which is determined by the information signal. An AM modulator for carrier frequency F_c and information signal S_t generates a modulated carrier according to the following equation:

$$AM = A \cdot [\cos (F_c) \times (1 + M \times S_t)],$$

where M is the *modulation index*, A is the final amplitude of the signal, and the information signal S_t varies between 0 and 1. M determines the extent to which the carrier signal is changed. The range of M is $0 < M < 1$, otherwise the carrier wave will be subject to disappearing ($S_t = -1$, $M = 1$) or abruptly changing phase ($S_t = -1$, $M > 1$), both of which have the effect of generating undesirable harmonics in the output signal.

When the signal is a cosine wave having frequency F_i, the equation above becomes:

$$\begin{aligned}
AM \quad &= A\cos (F_c) \times (1 + M \times \cos(F_i)), \\
&= A\cos (F_c) + M \cdot \cos (F_c) \cdot \cos (F_i), \\
&= A\cos (F_c) + \frac{A[M \cdot \cos (F_c + F_i)]}{2} + \frac{A[M \cdot \cos (F_c - F_i)]}{2}
\end{aligned}$$

This results in *three* spectral components: the carrier (F_c), the upper sideband ($F_c + F_i$), and the lower sideband ($F_c - F_i$). A time domain plot shows the modulated carrier; the "envelope" represents the information. A spectral plot clearly shows the carrier and the sidebands.

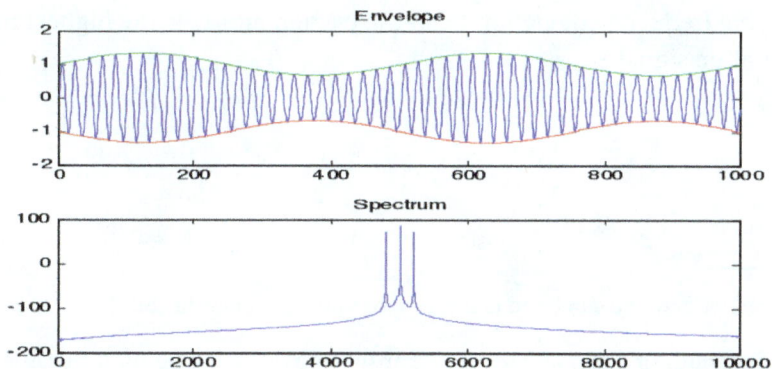

Figure 15. Amplitude Modulation example: time domain and spectrum.

For AM, the bandwidth of the modulated carrier is twice the bandwidth of the information signal, since AM generates both upper and lower side bands. The carrier frequency is in the center, and the sidebands are symmetric about the center frequency.

Single Sideband

Single Sideband (SSB) is a modulation technique that can be thought of as AM after filtering out the carrier and one of the sidebands[8]. If the upper sideband remains, the modulation is also called USB. If the lower sideband remains, the modulation is also called LSB. The spectrum of an SSB signal looks like the spectrum of the information signal, shifted up in the spectrum by the carrier frequency. An LSB signal is also spectrally reversed from the original information signal.

Because an SSB transmitter does not have to send the carrier and one of the sidebands, the bandwidth required is equal to the bandwidth of the information. Transmission is therefore more efficient.[9] SSB is also more complex (costly). Also, since SSB signals lack a carrier, off-tuning an SSB signal shifts all of the frequencies within the information signal, resulting in unusual effects. An SSB signal does not have a "center frequency", though we may still use that term to describe where to tune a receiver.

SSB is often used in the HF range by amateur radio operators. SSB CB transceivers are still available, at about twice the cost of normal (AM) CB transceivers.

Frequency Modulation

Frequency Modulation varies the carrier frequency in proportion to the information signal. An FM modulator generates a signal according to the following formula:

$$FM = \cos (F_c + MS_t),\text{[10]}$$

where M is again the modulation index, and is often expressed as F_Δ/F_s, where F_Δ is the maximum allowable frequency deviation from the carrier, and F_s is the highest frequency present within the information signal S_t.

---

[8] Filtering out the carrier and one sideband is one of two ways to generate SSB.

[9] The wider the bandwidth of a signal, the more transmit power must be used to raise it above the noise floor within potential receivers.

[10] This is actually phase modulation; frequency modulation uses the integral of the information signal.

46

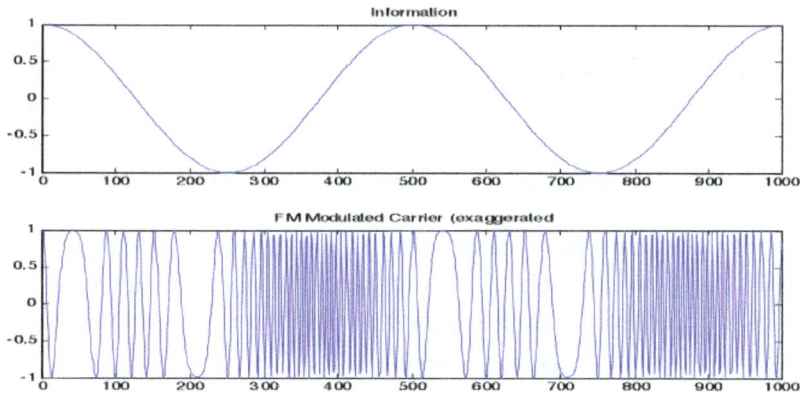

Figure 16. An information signal and the corresponding FM modulation of a carrier.

For a sinusoidal information signal, the resulting FM signal contains multiple frequencies. Most of the signal energy lies within a bandwidth of $2(F_\Delta + F_s)$. The frequencies present within the FM signal are highly redundant; the original information signal can be recovered if "most" of the frequencies are captured by a receiver.

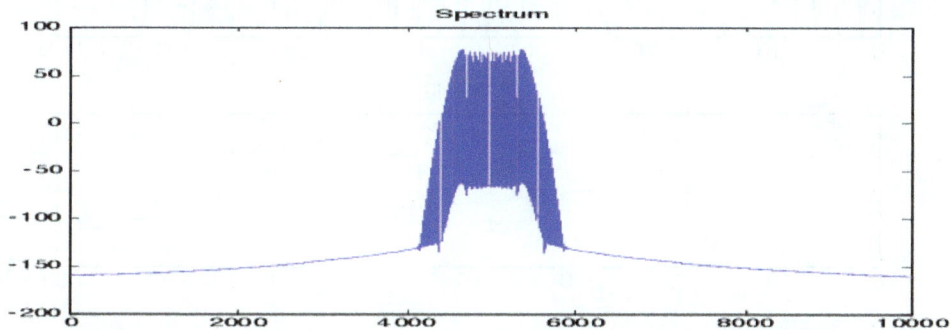

Figure 17. FM using a sine wave shows numerous spectral components.

An impressive feature of FM is its immunity to noise. This is due to the fact that FM is often transmitted at higher frequencies (where there is less background noise), and also the fact that FM receivers can actually eliminate amplitude variations during signal processing. This is accomplished through *limiting* the signal: the signal is amplified, then clipped (which limits the signal amplitude but introduces harmonics), and then filtered (to eliminate the harmonics), resulting in a nice FM signal having a uniform amplitude. FM receivers may have multiple limiting stages.

Frequency-Shift Keying

FSK is a digital modulation technique used within low BAUD-rate modems (Shanmugam [4]). We include it here because it is considered a form of FM. FSK uses one frequency to represent a zero, and a second frequency to represent a one. The frequencies must be higher than the baud rate so that the demodulator can differentiate between the 0 and the 1.

The spectrum of an FSK signal has two peaks, corresponding to the 0 and 1 frequency, assuming that the data signal has approximately as many zeros as ones. If the data signal contains predominantly zeroes, or predominantly ones, then the spectrum will show two peaks with one having a larger amount of signal energy than the other.

The bandwidth of an FSK signal must be wider than the distance between the two peaks. Higher baud rates require greater bandwidth.

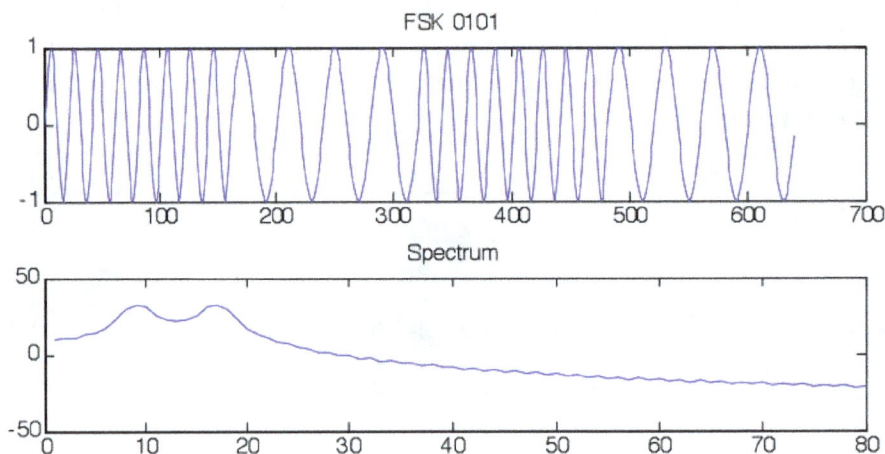

Figure 18. An FSK signal and its spectrum.

Other Modulation Types

A noteworthy analog modulation method is *vestigial sideband modulation,* named so because it partially suppresses one of the sidebands. This was used for analog television broadcasting. The wide bandwidth (4+ MHz) would require too much power if FM or AM were used, and SSB with its lack of a carrier frequency would have made proper tuning difficult. VSB keeps a little more than 1 MHz of the lower sideband and provides all of the upper sideband, along with a carrier reference. The total bandwidth requirement for analog TV is about 6 MHz.

3.12 Digitization (Sampling)

Digitizing a signal is accomplished by an *analog to digital converter* (ADC): a device which measures signal amplitude and provides a binary numerical output at fixed time intervals determined by a sample clock.

Critical operating parameters of an ADC include the maximum sample rate at which it can operate, the number of bits (resolution) that it provides, and the linearity of the result. ADCs providing a large number of bits (12 or more) and operating at high speeds (several megahertz) cost more than converters that operate at lower speeds or provide lower resolution.

Sampling must occur at precisely-timed intervals. Clocks that have significant phase jitter are not acceptable clock sources for ADCs.

An ADC *introduces* quantization noise. That is, a 12 bit converter can only represent 4096 discrete values. The analog input signal will be "between" values most of the time; the ADC will provide the binary value that is nearest to the signal level. The difference, which is ideally no more than one half of a bit of resolution for each sample, introduces noise into the digitized signal, since it is a departure from the ideal signal.

The output of an ADC will provide more accuracy when the input signal level matches the full input range of the ADC. When a low-level signal is applied, the ratio of the quantization noise to the signal level is high. When a high-level signal is applied, the ratio of quantization noise to the signal level is much lower.

The input signal level must not exceed the ADC's input range. A signal that exceeds the allowable input range cannot be represented within the limited number of output bits; the resulting binary value will be clipped, truncated, or grossly distorted in some other manner. Many ADCs have an *over range* output pin that indicates that the input exceeded the allowable input range.

3.13 More on Noise

Noise cannot be eliminated, but it can be managed. Within the digital domain, ensuring that quantities have enough bits, and that overflow is eliminated or carefully handled are necessary steps to manage noise.

Some processing, such as filtering, can actually reduce noise within the passband, if the noise meets certain random behavior criteria. Also, averaging techniques may be employed to reduce noise.

Demodulators are well understood, and textbooks on the subject indicate that for specific algorithms, an SNR above a specified level will guarantee a bit error rate below a specified threshold. After that, bit errors can often be eliminated by the use of error detection and correction encoding methods used in the data. However, even a strong error detection and correction method cannot compensate for one poorly implemented stage of a signal processing algorithm.

Exercises

1. The signals forming a dial tone are 350 Hz and 440 Hz. Other touch-tone signals also consist of two frequencies, which are not harmonically related, chosen from the table below (ITU [5]). What is the advantage to choosing signals that are not harmonically related?

	1209 Hz	1336 Hz	1477 Hz	1633 Hz
697 Hz	1	2	3	A
770 Hz	4	5	6	B
852 Hz	7	8	9	C
941 Hz	*	0	#	D

Table 5. DTMF ("Touch Tone") Frequencies

2. *A DTMF detector* is a circuit or a DSP algorithm that detects one of the touch-tone buttons above from the frequencies present within an input signal. Some wide-band signals, such as voices, modem signals, campaign speeches, or elevator music will have touch-tone frequencies present within them, such as 1209 Hz and 852 Hz (a '4' key). Describe a method that allows a detector to detect the DTMF signals, but reject wideband signals that contain DTMF frequency components.

3. A perfect A/D converter provides samples of 12 bits of unsigned integer values. What is the signal to noise ratio due to quantization? What is the SNR due to quantization error if the A/D converter performs signed conversion?

4. Describe the spectrum display for a touch-tone '6' signal.

5. A 770 Hz signal is subtracted from a 1477 Hz signal. Both have equal amplitudes. Describe the resulting spectrum.

6. A circuit is fed a 1000 Hz sine wave, and a 1.310 MHz sine wave, both having amplitude of 1.0 volts peak to peak. The output voltage of the circuit is described by the equation:

$$F(t) = A \cos(2\pi \times 1.310 \times 10^6 t)[1.0 + 0.25 \times \cos(2\pi \times 1000t)],$$

where A is 10.0, and t is time, in seconds.

   a. What is the peak to peak voltage of the output?

   b. Describe the spectrum of the output

c. Is this circuit a linear circuit? Explain.

7. Estimate the SNR of the spectra shown in Figure 9, Figure 10, and Figure 11.

References

[1] Chamberlin, Hal, *Musical Applications of Microprocessors*, Hayden Book Company, a division of Hayden Publishing Company, Inc., Hasbrouck Heights, New Jersey/Berkeley, California, 1985, pp. 14-15.

[2] Wikipedia, http://en.wikipedia.org/wiki/DBm.

[3] K. Sam Shanmugam, *Digital and Analog Communications Systems*, John Wiley and Sons, New York, 1979, pp. 251-304.

[4] K. Sam Shanmugam, pp. 408-413.

[5] International Telecommunications Union, ITU-T Recommendation Q.23, 1988.

Complex Arithmetic

4

In this chapter...

Complex numbers were often skipped in math classes; why bring them up now? The answer is that complex numbers and complex arithmetic simplify many engineering problems, especially where signal processing is involved. A single complex number can represent both the amplitude and the phase of a complex signal. The effects of multiplying and dividing signals using complex numbers is far easier than remembering laws for combining trigonometric functions.

4.1 Definitions

A *complex number* is a number that has both a *real part* and an *imaginary part*. If the imaginary part is zero, then the complex number is also called a *real number*.

Any operation applicable to a real number can also be applied to a complex number. In fact, the real numbers are a *special case* of complex numbers; complex arithmetic represents the more generic case.

Complex numbers simplify much of the math within signal processing. A single complex value represents a signal's amplitude and phase at a specific point in time.

It is an unfortunate artifact of history that imaginary numbers were called imaginary, implying that they have less credibility than real numbers. In fact, imaginary numbers are as "tangible" as real numbers. We are perhaps stymied because we don't use imaginary numbers when balancing our checkbooks (though for some people this may be difficult to tell).

The number i is number that, when multiplied by itself, produces -1. It is a constant, not a variable. Do not try and "figure out" what value it has. Do not try and look for it among the set of real numbers; it does not live on the real number line.

4.2 The Complex Plane

Complex numbers are written as a pair of numbers (real, imag) or sometimes as (real + i·imag). The complex number (3,-2) is also written as $3 - 2i$ or $3 - i2$, where i is the square root of -1. Engineering texts often use j instead of i, reserving i as a variable representing the current through a circuit.

While real numbers are visualized on a number line, complex numbers are visualized in a 2-dimensional plane, called the complex plane, or the *z-plane*.

Complex numbers are plotted on the complex plane, where the X axis is the real number line, and the Y axis is the imaginary axis. The plot below shows the complex point (3 – 2i). The real component of that point is 3, the imaginary component is -2.

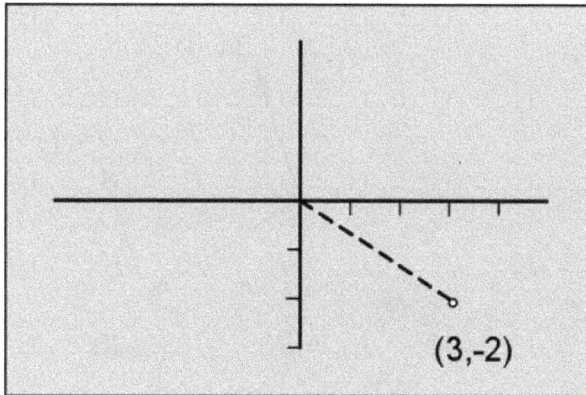

Figure 19. The complex number (3,-2) in the complex plane.

4.3 Operations

Adding and Subtracting Complex Numbers

Addition and subtraction are performed by adding (or subtracting) the real components, giving the real part of the result, and separately adding (or subtracting) the imaginary components, giving the imaginary part of the result. For example,

$$(3-2i) + (1+3i) = (3+1, -2i+3i) = (4+i)$$

This is similar to adding two polynomials:

$$(3-2x) + (1+3x) = (3+1 + -2x+3x) = (4+x)$$

The addition (or subtraction) of two complex values sometimes yields a real result:

$$(3-2i) + (1+2i) = (4+0i) = 4.$$

Complex addition can be visualized using *vector addition*: place the vectors head to tail and observe the new endpoint. In the example below, (3-2i) is added to (1+3i). The (1+3i) vector is shifted over to the point (3-2i), and the resulting new vector (in blue) points to the resulting complex value:

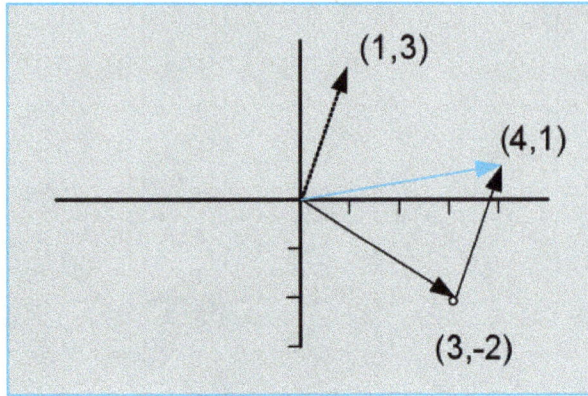

Figure 20. The sum of two complex numbers represented by vector sum.

Comparing Complex Numbers; Magnitude

Two complex numbers are equal if their real parts are equal and their imaginary parts are equal.

With real numbers, a real number $x > y$ if x is to the right of y on the real number line. This does not work with complex numbers, since they can be to the left or right of each other, and above or below each other.

Complex numbers are compared by comparing the *magnitude* of the complex numbers. This is the distance from the point (0, 0) to the complex number. This distance is always non-negative.

The real and imaginary components of a complex value form a right triangle with the X or Y axis. The magnitude is the length of the hypotenuse, which is computed by taking the square root of the sum of the squares of the real and imaginary parts. The magnitude of (3-2i) is therefore:

$$\text{sqrt} ((3 \times 3) + (-2) \times (-2))$$
$$= \quad \text{sqrt} (9 + 4)$$
$$= \quad \text{sqrt} (13).$$

The magnitude of z is often written as $|z|$. It is similar to the absolute value for real numbers. In fact, $|-3| = 3$, which is also abs(-3).

Several different complex numbers can have the same magnitude. For example, the complex numbers (1+0i), (-1+0i), (0-i), (0+i), and $(1/\sqrt{2} + i/\sqrt{2})$ all have a magnitude of 1. These points all lie on the *unit circle* within the complex plane; they have magnitude of 1.

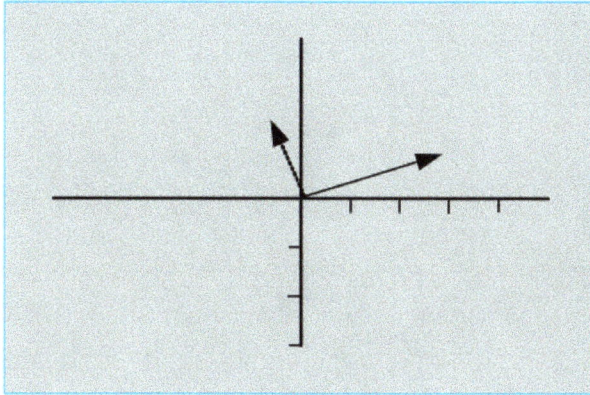

Figure 21. The "greater" of two complex numbers is the one farthest from (0,0).

Multiplication

Multiplication of two complex numbers is straightforward:

$$(a + ib)$$
$$\times \quad (c + id)$$
$$\overline{\hspace{4cm}}$$
$$(ac + iad + icb + i^2bd)$$

Recall that i^2 is -1. Using this, i^2bd becomes –bd. Then, we collect real and imaginary terms together :

$$(ac - bd) + i(ad + cb)$$

When b and d are zero, the result simplifies to ac, which is the product of the two real numbers (a + i0) and (c + i0).

Graphically, the product of two complex values yields a new vector: the length of the new vector is the product of the lengths of the multipliers, and the angle of the new vector is the sum of the angles of the two multipliers. (This is also true with real numbers; the angles are either 0 or $-\pi$).

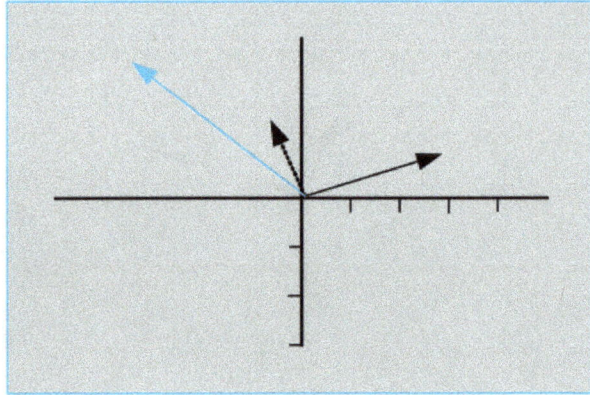

Figure 22. The product has the sum of the angles and product of the lengths.

Amches Pointer

Multiplication of two complex numbers having magnitude <= 1 yields a result with magnitude <= 1. Fractional binary 2s complement values can be used for real and imaginary components; multiplication should use overflow detection with saturation.

Conjugation

The *complex conjugate* of a complex value is the reflection about the X axis. The conjugate of a vector is computed by negating the imaginary component.

The complex conjugate of a real value is the original real value, since the negation of $0i$ is $0i$.

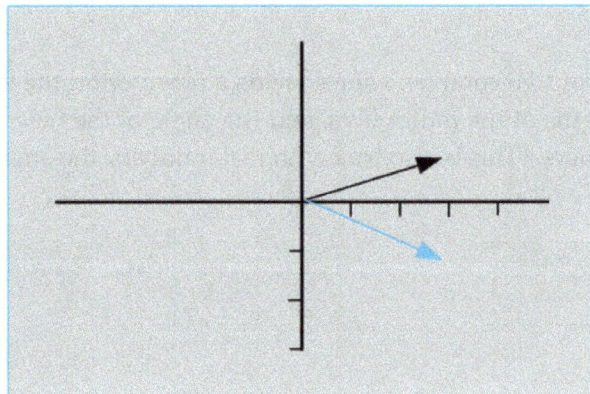

Figure 23. The conjugate is the reflection about the X-axis.

Any complex value times its conjugate value yields a real number. Graphically, since the conjugate is the reflection, and since the product of two complex numbers is the sum of the angles, the angle and its reflection cancel, leaving the result real axis.

Algebraically, if $z = (a + ib)$,

$$(a + ib)$$
$$\times \quad (a - ib)$$
$$\overline{\hphantom{-------------------------}}$$
$$(aa + \qquad iab - iab - i^2bb) = (a^2 - (-b^2)) = (a^2 + b^2) = |z|^2$$

For any complex value $z = (x + iy)$, $z \times conj(z)$ gives a real value $x^2 + y^2$. This value is the magnitude squared of the value; taking the square root gives the magnitude of the value.

Complex Division

Division is the inverse of multiplication[11]. Division by zero (where zero is $0 + 0i$) is invalid. When the divisor $(c + id)$ is nonzero, the quotient is computed as follows:

$$Q = \quad \frac{z_1}{z_2} \quad = \quad \frac{(a + ib)}{(c + id)}$$

It is undesirable and clumsy to have a complex value in the quotient. This can be addressed by multiplying the result by $(c - id)/(c - id)$, which is identically equal to one. Since $(c - id)$ is the conjugate of the denominator, the denominator becomes real after this multiplication:

$$Q = \frac{(a + ib)(c - id)}{(c + id)(c - id)} = \frac{(a + ib)(c - id)}{(c^2 + d^2)} = \frac{z_1 \times conj(z_2)}{|z_2|^2}$$

Since the denominator is now a real value, and since the numerator can be computed using the normal process for multiplication of two complex values, the result can be simplified to a complex value having no "i" term in the denominator.

Division undoes multiplication: within the complex plane, recall that multiplication of z_1 by z_2 results in a new vector, z_3, having a length equal to $|z_1| \times |z_2|$, and an angle equal to $angle(z_1) + angle(z_2)$. Dividing z_3 by z_1 gives z_2, where z_2 has a magnitude of $|z_3| / |z_1|$, and angle equal to $angle(z_3) - angle(z_1)$.

---

[11] Duh.

4.4 Polar Form

Complex numbers expressed as z = x +iy are said to be in rectangular form, since the location within the complex plane is at coordinates x, y.

The polar form of a complex number consists of a magnitude, normally called r (for radius), and an angle (usually in radians, normally called θ) with respect to the x axis.

Rectangular coordinates may be converted to polar coordinates using the following formulae:

$$r = \text{sqrt } (x^2 + y^2)$$

$$\theta = \tan^{-1} (y / x)$$

The arc tangent function must provide an angle placed in the proper quadrant; software functions use atan2 (y, x) to accomplish this. For example, the complex number (3,-2) has a radius of $\sqrt{13}$, and an angle of -0.5880 radians.

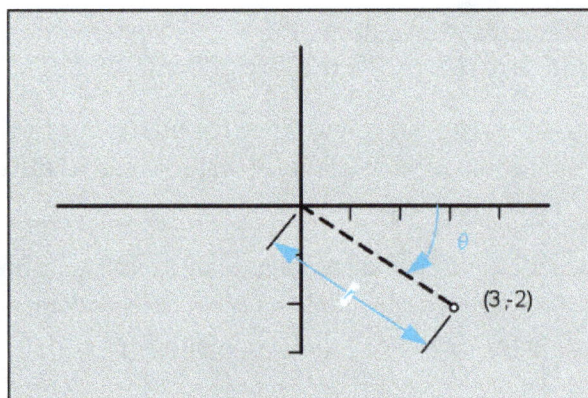

Figure 24. Polar form of (3, -2).

4.5 The Unit Circle

Within the complex plane, the *unit circle* is the set of all complex numbers having a magnitude of 1.0. These numbers have some special features that we can exploit:

- *Staying On the Unit Circle:* Since each number on the unit circle has a magnitude of 1, and since the product of two complex numbers yields a number whose magnitude is the product of the magnitudes of the multipliers, multiplication of z by any number on the unit circle gives a product having the same magnitude as z.

- *Incrementing Angles:* Since the product of two complex values gives a number having an angle which is the sum of the angles of the two multipliers, the resulting product will be rotated to a different position.

- *Complex Oscillator:* We can construct a *complex oscillator* by walking around the unit circle: start at (1,0), and, for each output point, multiply the previous point by (cos θ, sin θ). If $2\pi/\theta$ is an integer n, then we will end up back at (1,0) after n points. If $2\pi/\theta$ is not an integer, then after wrapping around the circle we will have a slight angular offset after n +1 points, which is OK.

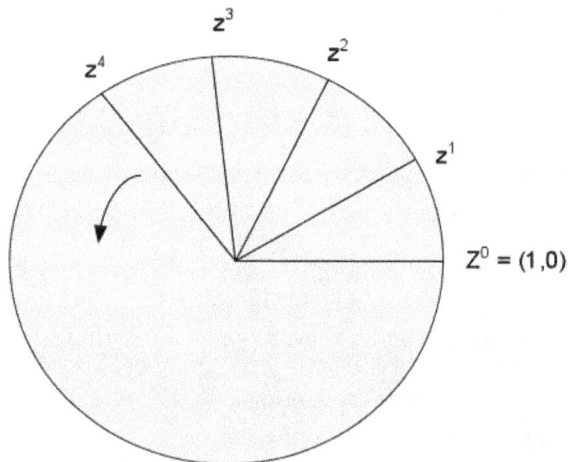

Figure 25. A complex oscillator formed by walking around the unit circle.

- *Real Oscillator:* A real oscillator may be obtained by taking either the real or imaginary result of a complex oscillator.

- *The n distinct n^{th} roots of 1:* While we know that 1 has two distinct square roots (1, and -1), it also has three distinct cubed roots. They are [1, (cos($2\pi/3$), sin($2\pi/3$)), (cos($4\pi/3$),

$\sin(4\pi/3))$]. These three points divide the unit circle into three equal wedges. There are also four distinct fourth roots of 1: $(1, -1, i,$ and $-i)$. Similarly, there are N distinct Nth roots of 1, the nth being located at $(\cos(2\pi n/N), \sin(2\pi n/N))$ (Churchill [1]).

If one constructs an oscillator using the above approach, beware that the values returned by *cos* and *sin* functions are approximations. Even double precision values will eventually "crawl off' the unit circle after some number of oscillator iterations. However, there are ways to correct for this.

Other Functions

It is possible to square, take square roots, compute logs and exponents, sines, cosines, etc. of complex values (Churchill [2]). When the imaginary part of the complex value is zero, the result reduces to a real operation.

Of special interest is complex exponentiation: $e^{(x + iy)}$. This is equivalent to $e^x \times e^{iy}$. This is also equivalent to: $e^x[\cos(y) + i\sin(y)]$, which is a computable complex number. This identity,

$$e^z = e^x[\cos(y) + i\sin(y)]$$

is known as *Euler's identity*, and is easily illustrated by taking the McLauren series for e^z and collecting the real and imaginary terms into separate infinite series (Kaplan [3]).

If y is zero, then $e^z = e^x$ and the result is real as would be expected when z is real.

Furthermore, if x is zero, the result is $\cos(y) + i\sin(y)$. Raising e to an imaginary value y yields the cosine of y plus $i \times$ sine of y. This gives rise to the fascinating identity:

$$e^{\pi i} = -1, \quad \text{or} \quad e^{\pi i} - 1 = 0,$$

which nicely ties together the transcendental numbers e and π with i, -1, and 0.

The expression $\cos(yt) + i\sin(yt)$ for some frequency y is often called a complex sinusoid. It is a sine wave having both real and imaginary components.

4.6 Phase

A single sinusoid is characterized by three parameters: its frequency, its amplitude, and its phase. Each of these quantities is meaningful when examined with respect to some reference:

- The frequency of a sinusoid is normally specified in cycles per second. However, in discrete time processing, the frequency of a signal must be expressed with the sample rate at which it was digitized.

- The amplitude can be specified in volts, millivolts, kilovolts, etc. However, within signal processing, it often makes sense to compare a signal level to the level of the noise floor, since signals at or below the noise floor will be of no value.[12]

- The phase of a signal is specified with respect to a fixed timing reference, such as the start of sample collection or some other timing reference.

A sinusoid represented as a set of complex values has a frequency, an amplitude (given by the magnitude of the complex value), and an inherent phase (given by the arctangent of the imaginary component divided by the real component). For a sinusoid of constant frequency, the amplitude and phase computed from any sample will not change from sample to sample.

Is Phase Important?

For sounds, slow changes in the amplitude or frequency of a signal are detectable. Slow changes in phase are not perceivable.

Why, then, should we care about phase? The answer is that systems can both *change* and *measure* phase. Therefore, the phase of a signal can be used to convey information, even though one cannot "hear" a change in phase. Consider this:

- When wiring two digital components together, the amplitude of signal lines between the components determines if the devices interpret the data as a logical 1 or 0. Binary data can be encoded as a voltage level.

- Phone lines, however, block DC voltages, so long strings of 0s or 1s would be heavily distorted or blocked entirely. Furthermore, a phone connection changes the amplitude of the signal between sender and receiver, and adds noise. Frequencies between 100 Hz and 3500 Hz can be passed from talker to listener. Even these frequencies will have some change in amplitude caused by the phone system, but they will pass through. Binary data can be encoded as different frequencies for 0 and 1 (*frequency shift keying*).

- Encoding the data using only two frequencies at a time leaves much of the spectrum available. Modems operating at 1200 BAUD or higher often employ phase shift keying (PSK): they vary the phase of the signal. When four phase values used, then it is quadrature PSK (QPSK). The receiver must measure phase with respect to some timing reference; it does this by detecting changes in phase from one bit time to the next, so the prior bit acts as a reference.

---

[12] Averaging methods and some filtering can improve the signal to noise ratio.

- This still leaves bandwidth available. Higher speed modems vary both the amplitude and phase of the signal. If there are N possible combinations of (amplitude, phase), this is called N-quadrature amplitude modulation (N-QAM). HDTV and satellite communications use 64-QAM or higher. A QAM signal requires a preamble that the receiver can detect; the preamble provides the amplitude and phase reference needed to determine the amplitude and phase of the signal that follows.

It is phase modulation and demodulation methods that facilitated the development of high speed modems, and high-bandwidth digital communications.

A Phase Demonstration

Look at the signals in the two figures below. Do you imagine that these would sound different? The first signal is a relatively nice square wave. The second signal looks very different.

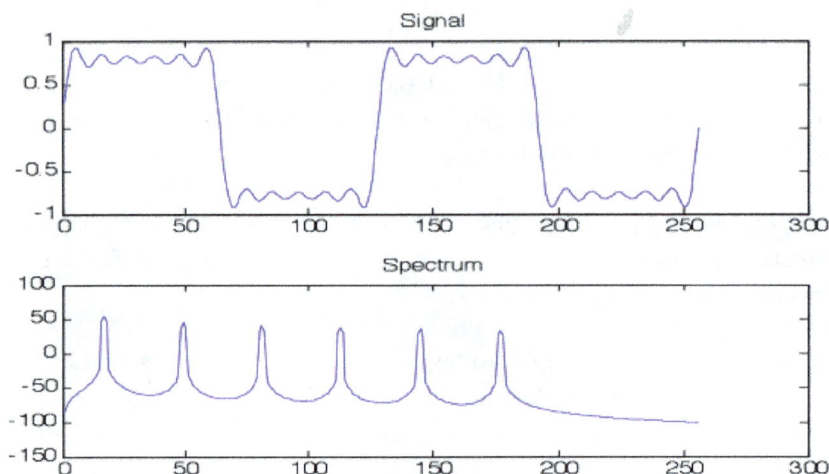

Figure 26. A relatively nice looking signal.

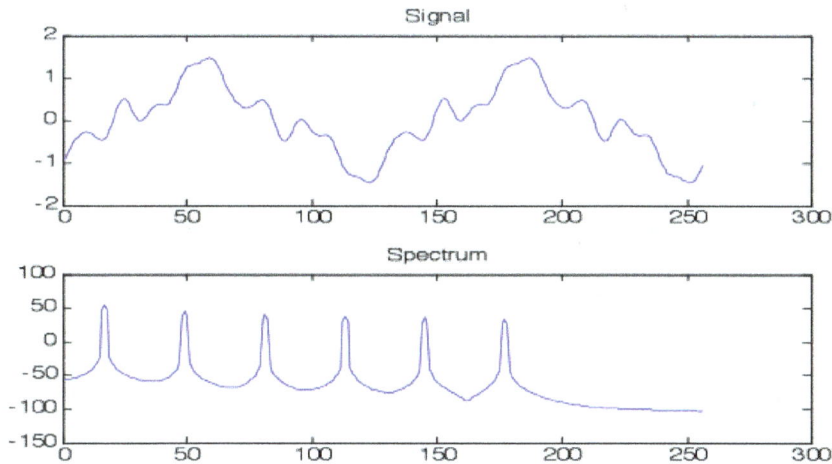

Figure 27. A messy signal having the same spectrum.

These signals, if played out of a speaker, would sound exactly the same. The second plot for each figure shows the spectrum; both signals have the same spectra. Indeed, the second signal was generated by randomizing the phase of the 6 frequency components that were used to generate the first plot. The signals have the same frequency components (which determine their sound) but different phases of those components (resulting in very different time domain plots).

4.7 Complex Arithmetic and Down Conversion

Down Conversion of a signal takes a signal that is centered around a frequency F_c and moves it "down" in the spectrum so that it is centered around 0 Hz. The result can be low-pass filtered (eliminating un-interesting signals in the spectrum). Once filtered, it is often *decimated* (reducing the sample rate), so that further operations require fewer operations per unit of time.

If the signal is centered at F_c hertz, then, in complex notation the signal is centered at $e^{2\pi i \times Fc \times t}$ where t is time in seconds. To down convert the signal to be centered at 0 Hz, we multiply each sample of the signal by a sample from the signal $e^{-2\pi i \times Fc \times t}$ (note the negative value):

$$e^{2\pi i \times Fc \times t} \times e^{-2\pi i \times Fc \times t} = e^{2\pi i \times Fc \times t - 2\pi i \times Fc \times t} = e^0 \text{, which is equal to the complex}$$
sinusoid of 0 Hertz.

Point by point multiplication of a signal with a complex sinusoid having a negative frequency rotates the spectrum of the signal left; point by point multiplication with a complex sinusoid having a positive frequency rotates the spectrum right.

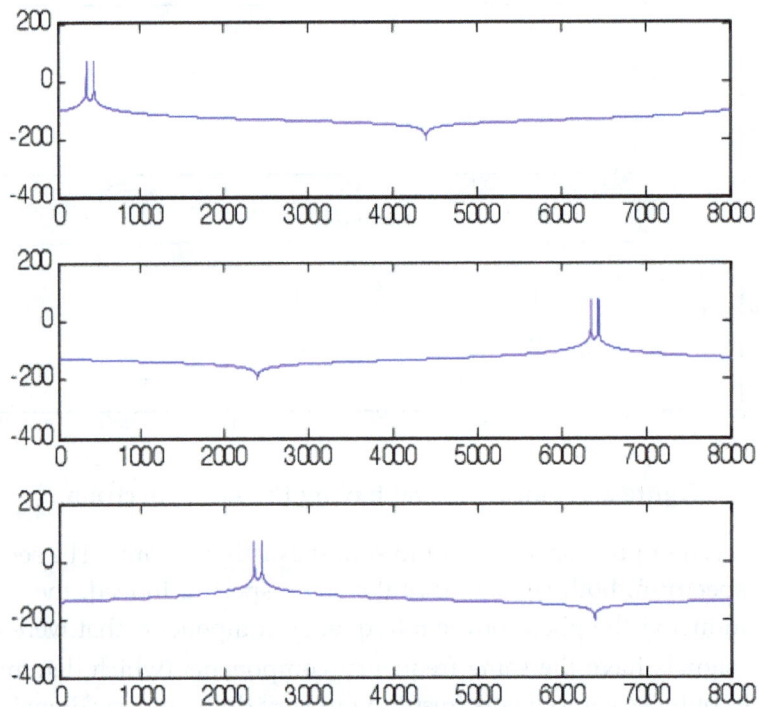

Figure 28. Original spectrum, rotated left, and rotated right.

Exercises

1. Re-order these complex values from smallest to largest: $(1 + 2i)$, $(1 - 2i)$, 1.0, -7

2. Compute the sum difference, and product of $(1+2i)$, $(3-i)$.

3. Raise the following complex values to the 8th power: 1, -1, i, -i, $(1+i)/\sqrt{2}$.

4. Geometrically, what happens when a complex value z is multiplied by its conjugate?

5. When any number is multiplied by 1, the result is the original number. What happens when any number is multiplied by a complex number having a magnitude of 1?

6. What might it mean to raise e to a complex value? What might it mean to take the natural log of a complex number?

7. Place two stereo speakers so that they face each other. Play your favorite music. Reverse the leads into one speaker. Is the resulting sound different? Explain.

8. A system provides 12 bit, signed, fractional complex input values to an FPGA. Write a VHDL routine to compute the magnitude squared of the input values, providing a 12 bit fractional result. What input values might cause overflow, and how should overflow be handled?

References

[1] Churchill, Brown, and Verhey, *Complex Variables and Applications,* Third Edition, McGraw-Hill Book Company, Hasbrouck Heights, New York, 1976, pp. 52-71.

[2] Churchill, p. 15.

[3] Kaplan, Wilfred, Applied Mathematics for Engineers, Addison-Wesley Publishing Company, Reading, Massachusetts, 1981, p. 108

Statistics

5

In this chapter...

When analyzing one or more data sets, it is often useful to summarize the data features using a few measurements. The mean and the standard deviation are often used, though there are multiple ways to summarize a data set.

By their nature, statistical measures sacrifice detailed information within the data set. It is possible to have two data sets that have the same mean and standard deviation but differ greatly on a point by point basis. Statistics should therefore be used with care.

Comparing two vectors measures the distance between corresponding points. If all points match, the vectors match. But often points may be "close" due to noise, distortions, and different gains and DC offset introduced by physical systems. The determination of whether two vectors match becomes a statistical assessment of the "distance" between two vectors.

5.1 Mean Value

The *mean* (μ) of a vector is the average of all of the values within a vector. The mean value of a complex vector is complex.

The mean value has a special interpretation for signals: it is the DC component of the signal.

5.2 Median Value

The *median* of a data set is the value in the "middle" of the set, when all values are sorted. The median of a complex data set is not defined, since complex values cannot be ordered like real values, though one could compute the median of the magnitude of all complex values in a signal. One could also compute the median of the real and imaginary components, and combine them into a single complex value. This value would live somewhere near the "middle" of the data set within the complex plane, though the point itself may not match any data points in the vector.

The median is useful when a data set has a distribution which has "outlying points" that adversely affect computation of the mean. For example, a process may generate a data set that contains a few extremely large positive or negative values, perhaps including some floating point positive or negative "infinity" values. The mean may be $+/-\infty$, or even non-a-number (NaN). However, the median of the data set will be reasonable as long as less than half of the data set consists of extreme values.

In practice, the median computation is easy to describe (sort the values, then choose the middle), but hard to compute, since sorts are time-consuming operations for large data sets.

Closely related to the mean are *percentile* values, where the list is sorted and the Nth value is chose, where N is (P/100)*length, and P is the percentile. For example, the 25th percentile value is determined by sorting the list and choosing the value that is at index (25/100)*length. The median is also the 50th percentile.

A common question arises when computing medians and the data set is such that there is no "middle" entry? For example, if a data set contains 11 elements, the 6th element is the center element, but if it contains 10 elements, there is no "center" element.

Approaches vary: since the median is not an exact parameter, choosing element 5 may be acceptable. Some implementations average elements 5 and 6 to produce the median.

A *median filter* should not be confused with the median. A median filter walks through a data set and computes the median value for points (0..N), (1..N+1), (2..N+2), etc. A median filter provides a new data set, each point of which is the median for a subset of data. The median provides a single value for a data set.

5.3 RMS

The root-mean-squared (RMS) value of a data set is the square root of the mean of the square of each point. For a complex data set, it is the square root of the mean of the magnitude squared of the data points[13]. The RMS of a complex data set is a real, non-negative value.

Like the mean, the RMS has a special meaning within electrical engineers. The RMS value of a time-varying voltage across a resistor is the DC voltage that yields the same average power as the time-varying voltage.

5.4 Maximum, Minimum

The *maximum* of a data set is the value of the largest value. If the vector is complex, the maximum is determined based on the magnitude of the complex values.

The *minimum* is the element having the smallest value. If the vector is real, this may be the most negative value. If the vector is complex, it is the value having the smallest magnitude.

5.5 Measures of Dispersion

The *range* is the maximum value minus the minimum value. The range is obviously affected by the extreme values within the data set, and can be wider than one would expect if the data contains a few extreme values. For a complex data set, the range can be defined as the complex value consisting of the individual ranges of the real and imaginary values. This defines the size of a "box" in the imaginary plane that can contain all points within the data set.

The *mean deviation* is the average distance between each point in a data set and the mean of the data set. This is a real value, even for complex data sets.

The *variance* and *standard deviation* are the most common measures of dispersion. These are both real numbers, even if the input data set is complex. The variance is the average of the squared distance between each point and the mean. The standard deviation is the square root of the variance.

Algorithmically, the standard deviation of a real vector is computed as follows:

```
/** compute the mean **/
for (mean = 0, j = 0; j <= nPoints; j += 1)
        mean = mean + vector[j];
mean = mean / nPoints;
```

---

[13] Remember that the magnitude squared of a complex value can be computed by multiplying the value by its conjugate value.

```
/** sum square of distance from each point to the mean **/
for (s2 = 0, j = 0; j <= nPoints; j++)
        s2 = s2 + (vector[j] - mean) * (vector[j] - mean);


s2 = s2 / nPoints;      /** average value of s2 = variance **/

stdDev = sqrt (s2);     /** standard deviation  **/
```

For a complex vector, the standard deviation is computed in a similar fashion, except (1) the mean of the vector is complex, and (2) the sum of the distance squared is the sum of the real valued distance squared between each point and the mean for each point.

In the complex plane, the mean is the "center point" of the data set, and the standard deviation is a radius about the mean into which most points in the data set fall.

Amches Pointer

The mean value of a signal is equivalent to the DC value. The standard deviation is equivalent to the RMS of the AC component of a signal. The range is equivalent to the peak-to-peak voltage of a signal.

5.6 Comparing Vectors: Mean Error

Two digitized sets of vectors "look similar" if corresponding points have approximately the same values. If, for example, we have real data on the value of a stock, and we also have a mathematical model of the stock value, we may wish to determine how similar the two are before making a stock transaction based on what the model predicts.

One approach to measure the similarity would be to subtract corresponding points, take the absolute value of the distance, sum these, and divide by the number of points. This would give the average error per point.

This is the mean error, given by:

$$\text{Mean error } (X, Y) = (1/N) \sum | x_i - y_i |$$

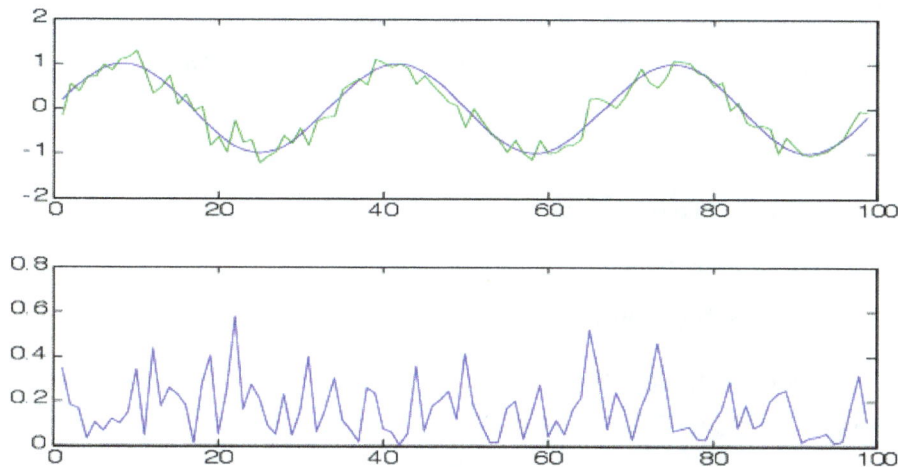

Figure 29. Two vectors and the absolute value of their difference.

In Figure 29, the mean error is 0.163. This value could be compared to some threshold to determine if the vectors are "close enough". Of course, if the vectors are larger (say, 10x larger), then the mean error would also be 10x larger, or 16.3.

5.7 Comparing: RMS Error

The RMS error is more widely used to determine the distance between two vectors. The RMS error for the two vectors in Figure 29 is 0.205. The RMS error computes the RMS of the point-by-point difference between two vectors.

For the figure above, the RMS error is larger than the mean error. The RMS error is affected more significantly than the mean error if the error vector contains a small number of large values.

For example, suppose two 10-element vectors have differences of (10,1,1,1,1,1,1,1,1,1). The mean error is 1.9; the RMS error is 3.3.

Amches Pointer

The standard deviation is the RMS error between a data set and its mean value. This is also the RMS value of the AC components of the signal.

73

5.8 The Dot Product

The *dot product* is a useful scalar value computed from two vectors. The dot product is central to these computations:

- The RMS value of a set of samples
- The correlation of two sets of samples
- The Discrete Fourier transform of a set of samples
- The convolution of two sets of samples

The many applications of the dot product justify the multiply-accumulate instructions (MAC) present in many digital signal processors.

Dot Product Definition

You may recall the dot product from physics has the following characteristics:

- For two vectors x, y, the dot product x·y is a scalar value
- For vectors in a plane that go from the origin to (x_1, y_1) and (x_2, y_2), the dot product is computed as follows:

$$x \cdot y = x_1 x_2 + y_1 y_2$$

- It is also computed using $|x| |y| \times \cos(\theta)$, where θ is the angle between x, y.
- It is zero if the vectors are at right angles to each other. Vectors at right angles are said to be *orthogonal*. Changes to one vector do not affect the other.
- It is at a maximum positive or negative value if the vectors point in the same (or in exact opposite) directions
- The dot product of a vector with itself gives a maximum positive value equal to the length squared of the vector.
- It is sometimes called the "projection" of one vector onto another. That is, loosely, it represents a measure of how much one vector is pointing in the same direction (or opposite direction) as the other vector.

Vector Dot Product

The dot product in physics operates on vectors living within two or three dimensional space. We now turn our attention to the dot product of two vectors, where the vectors are now arrays of *n* values. The dot product has these attributes:

- For two vectors x, y, x·y is a scalar value

- It is computed from the sum of multiplying corresponding terms in x, y

- It is zero if the vectors are *orthogonal* to each other

- It is at a maximum positive or negative value if the vectors are co-linear within an n-dimensional space.

- The dot product of a vector with itself gives a maximum positive value.

- It is sometimes called the "projection" of one vector onto another. That is, loosely, it represents a measure of how much one vector is pointing in the same direction (or opposite direction) as the other vector within the n-dimensional space.

The dot product of two vectors X, Y is computed using the following:

$$X \cdot Y = \sum (x_i\, y_i)$$

Corresponding elements from each vector are multiplied together, and the resulting products are added into a single value.

If Y is a complex vector, then the dot product uses the conjugate value of Y. If X, Y or both vectors are complex, then the resulting dot product is also complex.

The RMS can be redefined using the dot product of a vector with itself. In this case, if X is complex, then X·X will be real, because the dot product uses the conjugate of the second vector, and vector elements times their conjugates are always non-negative and real.

$$RMS(X) = sqrt\ [(X \cdot X)\ /\ length(X)]$$

The variance and standard deviation can be expressed in terms of the dot product:

$$var(X) \quad = [(X - \mu) \cdot (X - \mu)\ /\ length(X)]$$

$$stddev(X) = sqrt\ [(X - \mu) \cdot (X - \mu)\ /\ length(X)]$$

5.9 Measuring Similarity Using the Dot Product

The dot product can be used to compare two vectors. The dot product provides more information than the RMS error or mean error.

The dot product can test the similarity of two vectors having a mean value of zero. (The mean can be subtracted from each element of a vector in order to produce a vector with zero mean).

The dot product between two vectors having zero mean will be a maximum positive value if the vectors have the same "shape". If one vector is the negative of the other (that is, one vector increases while the other decreases, and vice versa), then the dot product will generate a large negative value. If the two vectors have dissimilar shapes then the dot product will be closer to 0.

The magnitude of the dot product is determined by the "largeness" of the values within the two vectors. If we want to compare the "shapes" of two vectors but ignore their relative magnitudes and DC offsets, each vector should first be "normalized" by 1) subtracting the mean value from the vector, and 2) dividing each vector by the standard deviation. Normalization does not affect the shape of the vector.

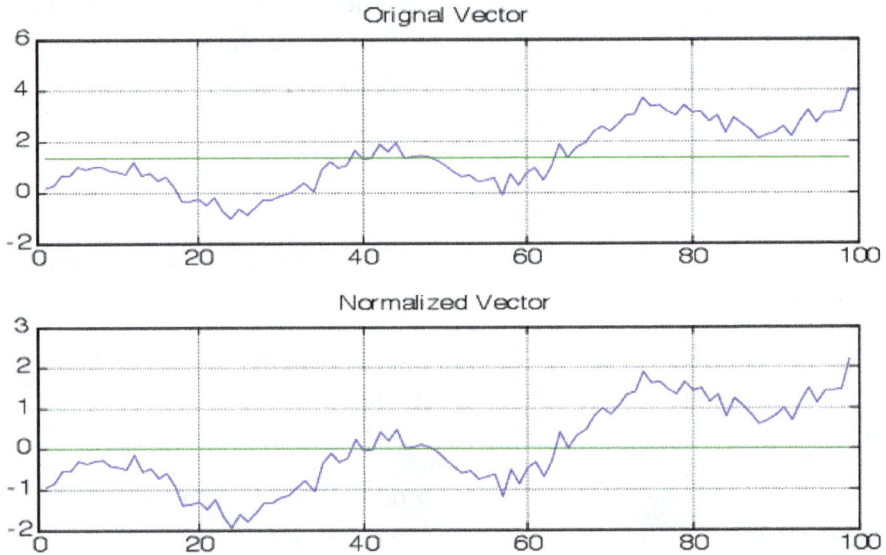

Figure 30. A normalized vector having 0 mean and standard deviation = 1.

Exercises

1. Compute the dot product of the vector (1, 2, 3) with
   - a. (1, 2, 3)
   - b. (2, 3, 1)
   - c. (3, 1, 2)

2. Compute the dot product of the vector (1, 2, 3) with
   - d. (2, 4, 6)
   - e. (4, 6, 2)
   - f. (6, 2, 4)

3. Compute the dot product of each vector in exercise 2 with itself.

4. A single cycle of a cosine, sampled 8 times, has samples:
   $$\{1, \quad 0.7071, \quad 0, \quad -0.7071, \quad -1, \quad -0.7071, \quad 0, \quad 0.7071\}.$$
   A cosine having twice the frequency has samples:
   $$\{1, 0, -1, 0, 1, 0, -1, 0\}.$$
   Compute the dot product of these two vectors.

5. A cosine of frequency 0 Hz has samples {1,1,1,1,1,1,1,1}. Compute the dot product of this with the two vectors provided in exercise 4.

6. When might one prefer to use the median value of a set of samples instead of the mean value?

7. The mean value of a set of voltage samples is high, while the variance is very low, but nonzero. What does this say about the signal?

8. The variance of a set of voltage samples is high, but the mean is low. What does this say about the signal?

9. A signal has a high mean value and a high variance. What does the signal look like?

10. Two signals have the same mean value and the same variance. Are the signals alike? Explain.

Correlation and Convolution

6

In this chapter...

Correlation is the process of determining how closely two vectors match. *Convolution* is closely related, and is often used to filter data within the time domain. Both correlation and convolution are based on the dot product.

Correlation can be used to find the occurrence of a pattern within a data set. Pattern matching like this may be used to identify a preamble signal by matching the spectrum of incoming data with the pre-computed spectrum of the preamble. The incoming data may have noise or varying amplitudes based on the transmission media; correlation allows us to find a "best match" even when the input signal is imperfect.

Introduction

Correlation is the process of determining how closely two vectors match. *Convolution* is closely related, and is often used to filter data within the time domain. Both correlation and convolution are based on the dot product.

Correlation can be used to find the occurrence of a pattern within a data set. Pattern matching like this may be used to identify a preamble signal by matching the spectrum of incoming data with the pre-computed spectrum of the preamble. The incoming data may have noise or varying amplitudes based on the transmission media; correlation allows us to find a "best match" even when the input signal is imperfect.

Convolution is closely related, but is often used to implement filters.

Correlation uses the dot product to measure the similarity between two vectors.

The dot product between two vectors will be "larger" if the two vectors have "similar" shapes. It will be a large negative value if one vector is similar to the negative value of the other vector. If the two vectors are dissimilar, the dot product will be closer to zero.

6.1 Distance Measures and Pattern Matching

Consider the following real-world problems:

- A modem examines a sample stream of data, looking for a specific *preamble* which indicates data is to follow.

- A known signal has a specific spectral "shape". A system looks for this signal by examining data in the frequency domain.

- A self-test function generates a test signal out; we want to determine how closely the test output matches the ideal results in order to determine if the system is operating correctly.

These are examples where it is necessary measure the degree to which two vectors are similar. In each case, there is input data (a sample stream, FFT results, etc), and a pattern which we seek to find within the input data. The input data may have no signal (if the data source is not active), it may have high levels of noise (if the channel is noisy), it may have some "coloration" (that is, uneven filtering of some frequencies due to channel limitations), and it may have an unpredictable amplitude (due to random gain or attenuation in antennas, amplifiers, ADCs, etc). Our goal is to determine if the signal contains the desired pattern.

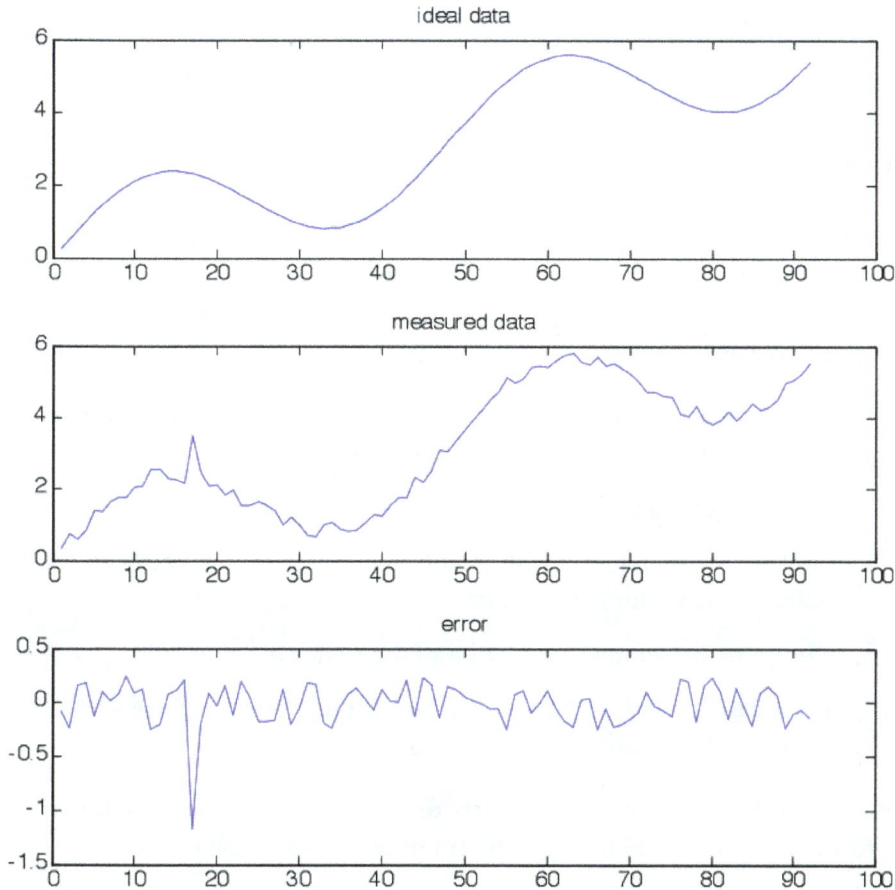

Figure 31. Errors between a clean signal and a noisy version.

Figure 31 illustrates an ideal signal, a measured signal, and the error (the point to point distance) between the ideal signal and the measured signal. The mean error is 0.1384, while the RMS error is 0.1888.

6.2 Correlation Coefficient

Formulas for computing distance work well when comparing vectors having the same DC offset and amplitude range. Suppose that vectors do not have the same DC offset or magnitude, but rather the same "shape". For example, the model of a physical system may generate a vector having a range of values between 1 and -1, but the physical system, due to use of an 8-bit analog to digital converter, generates values between -128 and 127. In this case, mean and RMS distance measures are not helpful.

The correlation coefficient is applicable here. When two vectors are well-correlated, the correlation coefficient will be close to 1.0. This indicates that both vectors have approximately the same "shape" – they have similar bumps in similar places. The vectors may have different scales, but they look similar. If the correlation coefficient is close to -1, this indicates inverse correlation. Inverse correlation occurs when one vector has the same shape as the negative of the other vector. A correlation coefficient near zero indicates the vectors are poorly correlation: the shape of one vector does not correspond to the shape of the other vector. Figure 32 illustrates these cases. Scaling for the vectors varies greatly, yet the correlation coefficient is always in the range -1 to 1.

The correlation coefficient is computed by first "normalizing" the two vectors. This process centers each vector and scales it to a standard range. The following steps normalize a vector:

1. Compute the mean of the vector μ.

2. Subtract the mean from each point. This "centers" the vector so that the average value of the vector is now zero.

3. Compute the standard deviation of the vector σ.

4. Divide each point by the standard deviation. This gives the vector a "unit standard deviation."

The correlation coefficient is the dot product, divided by the length, of the two normalized vectors. The formula for the correlation coefficient of two un-normalized vectors X, Y is:

$$r = \frac{1}{N} \times \frac{(X - \mu_x) \cdot (Y - \mu_y)}{\sigma_x \sigma_y}$$

When correlating against a fixed pattern, the pattern can be "pre-normalized" and placed in the Y vector, so that the correlation coefficient becomes:

$$r = \frac{1}{N} \times \frac{(X - \mu_x) \cdot Y}{\sigma_x}$$

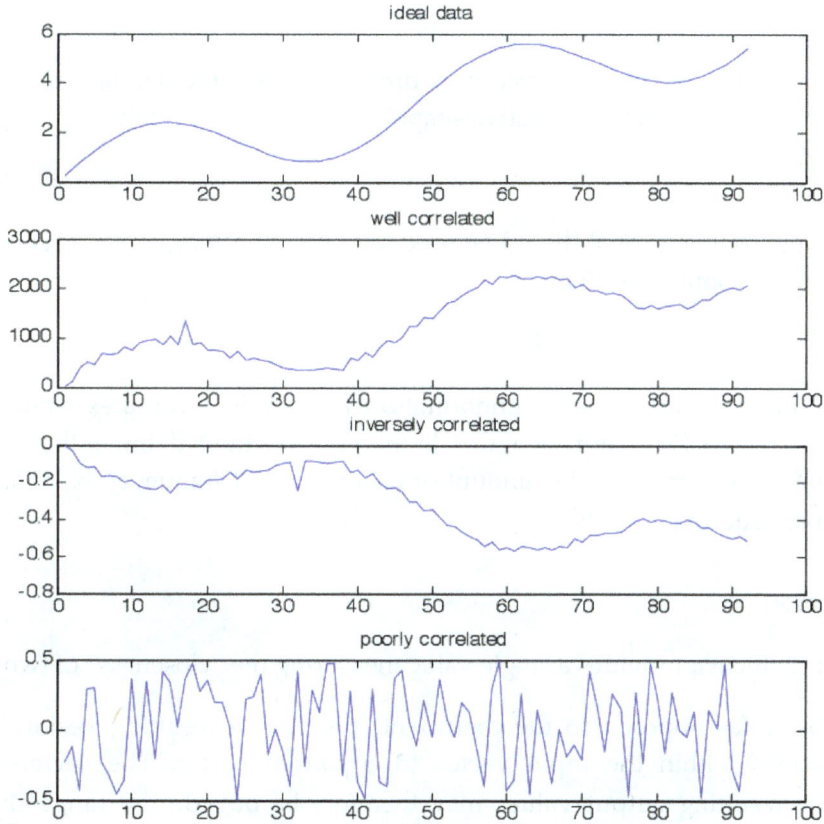

Figure 32. Correlation and Inverse Correlation.

In Figure 32, the first two vectors have the same "shape", but the amplitude of the second vector is huge compared to the pattern. The vectors have the following characteristics:

Vector	Mean	Std Dev	Correlation w/ Ideal
ideal	3.1126	1.6702	1.0
well-correlated	62.95	33.7535	0.9944
inversely correlated	-0.3151	0.1676	-0.9912
poorly correlated	-0.220	0.3009	-0.0596

Even though the first two vectors have very different ranges, their correlation coefficient is 0.9944; they are well-correlated.

The inversely correlated vector also has a very different mean and standard deviation from the ideal vector, but because it has the "negative shape", it correlates inversely.

Complex Correlation

If Y is complex, the conjugate of Y should be used, and r will be complex. This is unusual, but an interesting and more common variant is:

$$r = X \cdot Y$$

where X is an input vector (possibly unnormalized) and Y is a complex sinusoid having an integral number of cycles (N) over its length[14]. In this case, r is the Nth bin of the FFT of the signal X. The magnitude of r represents the amount of energy in X at frequency N, and the angle of r represents the phase at frequency N.

6.3 Correlation

The correlation coefficient provides a single value measuring the "closeness" of two vectors.

General correlation (as opposed to the coefficient) provides a resulting vector, by comparing successive locations within the input vector to a pattern vector. The vectors need not be normalized. The resulting output values may therefore be outside the range -1 to 1. The nth output value is computed from:

$$W_n = \sum (x_{i+n} \cdot y_i)$$

The result W_n is the dot product of the pattern Y with a vector X starting at offset n within the vector. When searching for a pattern within a stream of samples, this unnormalized correlation saves time. Instead of looking for a value near 1, indicating a good correlation, we simply look for a *maximum* value, indicating the best correlation within a data set.

Consider the case where we want to "search" a noisy vector for a specific sub-vector pattern. In this case, we want to correlate the sub-vector with each section of the vector we are searching, and find that portion of the vector that best matches the sub-vector pattern:

---

[14] A complex sinusoid having an integer number of cycles has mean 0 and standard deviation 1; it is therefore normalized.

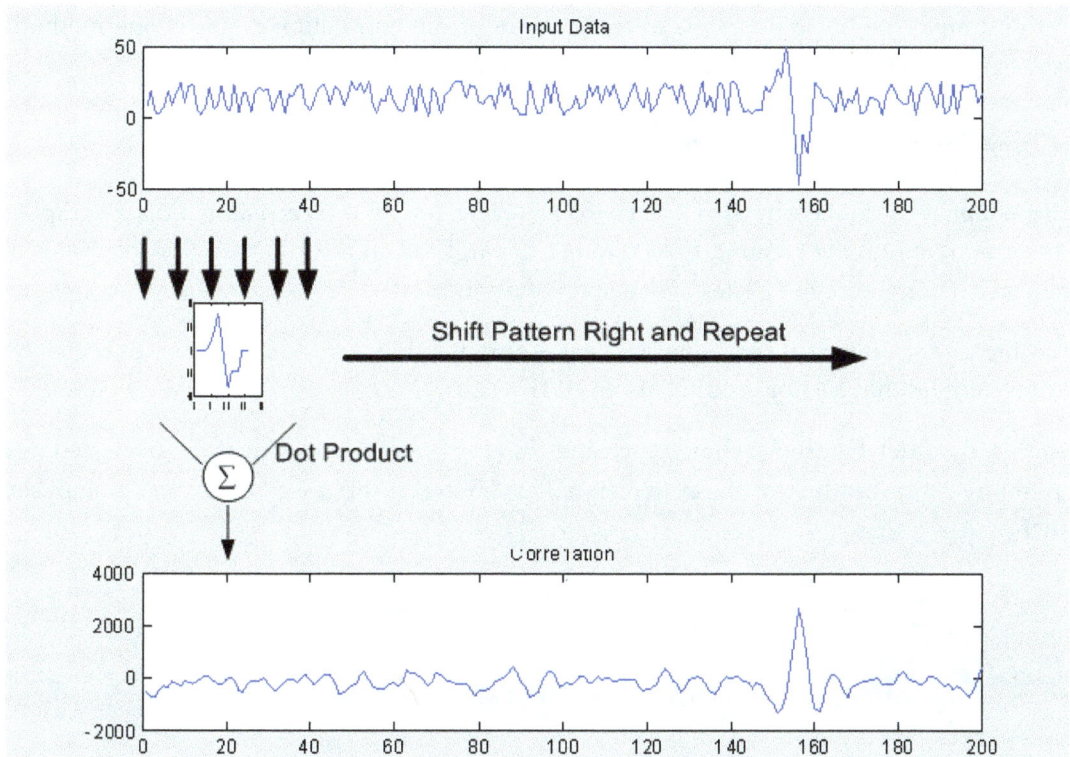

Figure 33. Correlation against a pattern.

In this case, the pattern is 20 samples wide, and the input data is 200 samples wide. Finding the pattern is accomplished using these steps:

1. Select 20 samples from the input vector: samples n to n+19.

2. Compute the dot product of the selected samples and the pattern; this becomes the next output sample.

3. Increment n, and repeat.

The output sample having the largest value represents the best match; the index of this peak corresponds to the location of the best match within the input vector. Pattern matching is therefore accomplished by computing a series of dot products and finding the maximum result. The index of the peak corresponds to the index of the samples that best match the pattern.

When the input vector is all noise, the correlation will produce a maximum value *somewhere*. The question becomes, is the peak high enough to represent a real signal? This is normally determined by requiring that the peak be some minimum level above the noise. There are several ways to estimate the noise floor: the average value of the correlation result is one method; the average excluding the most prominent N peaks is another method.

6.4 Correlation of Frequencies

A real frequency is defined by $s_t = \cos(2\pi ft/F_s)$. Here f is an integer value, F_s is a sample rate, and t is a sample number ranging from 0 to n-1. If $f \cdot n/Fs$ is an integer, then this generates a real cosine wave having exactly f cycles over a span of n samples.

Two cosine signals, one at f1, and one at f2, both of which have an exact number of cycles over a span of n samples, have a dot product of zero if f1 ≠ f2.

To see this, consider Figure 34, showing cosines having 5 and 7 cycles over the indicated period. The point-by-point product of these two signals is shown in Figure 35. By way of analysis, the product is:

$$s_t = \frac{\cos(2\pi(f_1 + f_2)t/F_s) \ + \cos(2\pi(f_1 - f_2)t/F_s)}{2}$$

$$= \frac{\cos(2\pi \cdot 12t/F_s) \ + \cos(2\pi \cdot 2t/F_s)}{2}$$

The product of the two cosines is the sum of two cosines of different frequencies, 2 cycles and 12 cycles, scaled to ½ the original amplitude. The sum of each of these cosines, computed over 256, is zero, since the cosines have an exact number of cycles.

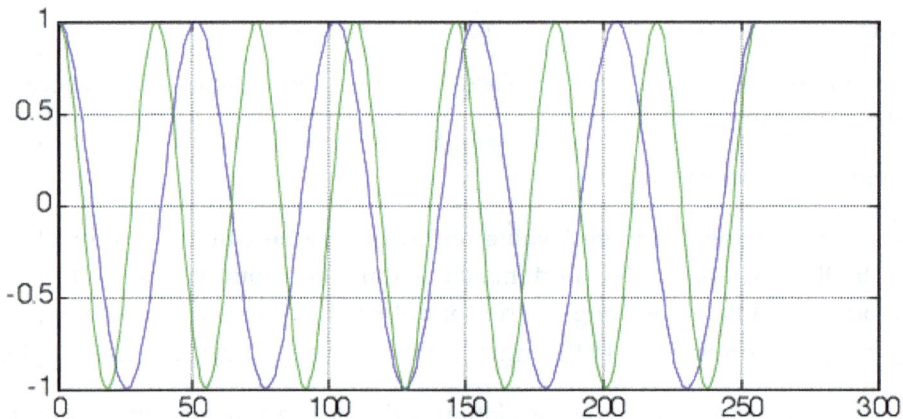

Figure 34. Cosines of 5, 7 cycles over 256 samples.

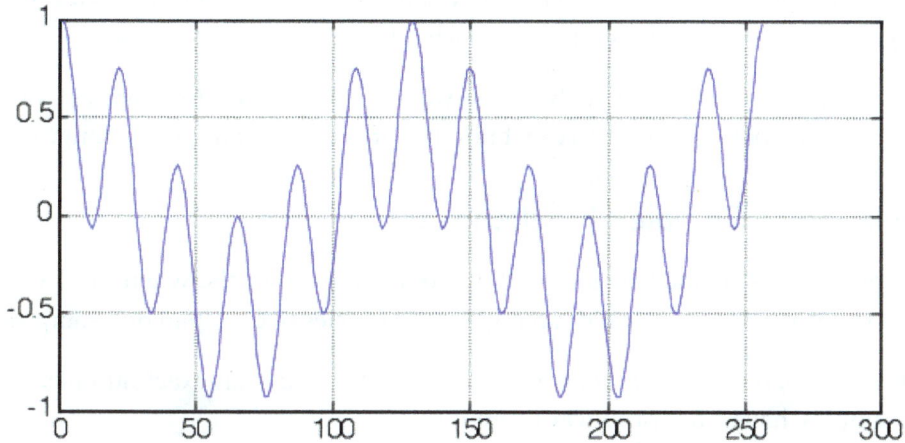

Figure 35. Product of cosines of 5, 7 cycles.

If, however, f1 = f2, as when a cosine of 5 cycles is point-by-point multiplied with itself, the result is given by the equation:

$$s_t \; = \; \frac{\cos(2\pi \cdot 10t/F_s) \; + \cos(0)}{2} \; = \; \frac{\cos(2\pi \cdot 10t/F_s) \; + 1}{2}$$

This is a cosine of frequency 10 and amplitude ½ plus a scalar of ½. Summing these terms over all points over all points in the vector causes the cosine term to vanish, leaving the result as n/2, where n is the vector length.

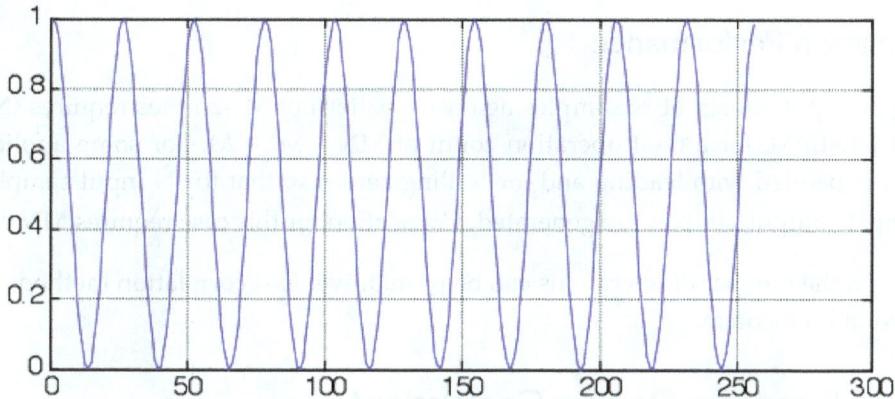

Figure 36. Product of cosine of 5 cycles with itself.

Therefore, cosines of different frequencies are orthogonal with each other, provided the cosines generate an exact number of cycles over the same sample range. A similar approach shows:

- Sine waves of different frequencies, generating an exact number of cycles over a specific range of samples, are orthogonal with each other.

- Sine waves are orthogonal with cosine waves, whether the frequencies are different or not, provided both generate an exact number of cycles over a specific range of samples.

6.5 Correlation of Complex Frequencies

A complex frequency is defined by $s_t = e^{2\pi jft/Fs}$. As with real signals, when f·n/Fs is an integer, then this generates a complex sinusoid having exactly f cycles over a span of n samples.

Two complex signals, one at f1, and one at f2, both of which have an exact number of cycles over a span of n samples, have a dot product of zero if f1 ≠ f2.

$$s_t = (e^{2\pi jf1t/Fs})(e^{-2\pi jf2t/Fs}) \; = \; e^{2\pi j(f1-f2)t/Fs}$$

Here we use the conjugate of the second frequency. This is a new sinusoid, having a frequency of f1-f2, which also has an exact number of cycles over a span of n points. The sum of all points in this span will be zero.

However, when f1 = f2, the term-by-term product is

$$s_t = (e^{2\pi jf1t/Fs})(e^{-2\pi jf1t/Fs}) \; = \; e^{2\pi j(0)t/Fs} = 1.$$

As with real signals, the dot product of two complex sinusoids with different frequencies is 0; the dot product of a complex sinusoid with itself is n, where n is the length of the vector.

6.6 Correlation Performance

Correlating an input stream of N samples against a pattern of M samples requires (N-M+1) dot products of length M, for a total operation count of $MN - M^2 + M$. For some applications, the input vector is padded with leading and/or trailing zeroes so that for N input samples plus the zero padding, N output samples are generated. Correlation for this case requires MN operations.

For large M, N, the number of operations can be prohibitive. Fast correlation methods exist based on the convolution theorem.

6.7 Time or Frequency Domain Correlation?

Time domain signals can be correlated, and spectra can be correlated. In either case, a higher value means a closer match.

If a system does not distort phase, and if phase relationships *are* critical to pattern identification, time domain correlation is applicable. If phase relationships cannot be guaranteed, or if they are irrelevant, then the magnitude of the FFTs of the signal and pattern can be correlated; this eliminates all phase information.

6.8 Convolution

The mathematical operation called *convolution* is related to correlation:

$$W_n = \sum (x_{i+n} \cdot y_{m-i})$$

This is the correlation formula, except that the pattern, y, is reverse ordered in the dot product. Reversal is denoted by using the m-i subscript for the Y pattern vector.

A FIR filter computes the convolution of an input sample stream with a set of numbers known as the FIR filter *taps* or *coefficients*.

Experienced developers may say, "but wait. Every FIR filter I've seen has coefficients that are the same when reversed, such as {2, 4, 5.3 6.1, 5.3, 4, 2}". It is true that these filter taps are *symmetric*, and therefore reversing them makes no difference. However, filter coefficients need not be symmetric. Also, convolution is used for more than FIR filters. It is therefore important that we remember the strict definition of convolution.

Amches Pointer

Most FIR filters have symmetric taps; symmetric taps cause the filter to be phase linear. But not all FIR filters require linear phase, and therefore not all FIR filters have symmetric taps.

A FIR filter may be thought of as a series of dot products between an input vector and the reversed filter taps. In this sense, a FIR filter can also be thought of as measuring how well the input signal stream correlates to the reversed taps of the filter.

6.9 The Convolution Theorem

The Convolution Theorem states that convolution of two signals in the time domain is equal to the inverse FFT of the point-by-point product of the FFTs of the two signals (Lyons [1]). In other words, an alternate way to compute the convolution of two signals is as follows:

- Compute the FFT of each signal

- Compute the product, point by point, of the two FFTs

- Compute the inverse FFT of this signal

The number of operations needed to compute the convolution between two vectors is proportional to the *product* of the *lengths* of the vectors. However, for long vectors, it is more economical to use the above approach.

Example: the correlation of two 1024 point complex vectors requires more than 6 million operations. However, the FFT of a 1024 point complex vector requires about 50,000 operations. The inverse FFT, and the two FFTs require 150,000 operations. The point by point multiply requires 6000 operations. The convolution theorem requires 156,000 operations, versus 6 million operations, an improvement of about 40 times for complex vectors of length 1024.

There are some caveats to this approach:

- One must remember that the FFT and inverse FFT operate on the periodic extension of the signals.

- The vectors must be of the same length; zero padding is often employed to ensure this.

Exercises

1. Describe how to use correlation to determine if an input signal contains a dial tone (350, 440 Hz).

2. What might you expect if a random set of samples is correlated against a random signal?

3. The *autocorrelation* of a signal is the result of correlating a signal with itself. Explain what might cause peak values in the autocorrelation result.

References

[1] Lyons, Richard G., *Understanding Digital Signal Processing*, Second Edition, Prentice Hall, Upper Saddle River, New Jersey, 2004, pp. 200-207.

More on Frequencies

7

In this chapter...

We've already referenced frequencies several times, but we've not addressed the differences between real and complex frequencies. In this section, we revisit frequencies, and provide more detail on the relationships between real and complex frequencies, and their uses.

7.1 Real Sinusoids

A real sinusoid is defined by three parameters: the amplitude (that is, how "high" is the signal), the frequency (usually in cycles per second), and the phase (that is, delay: where does the signal begin with respect to some time reference).

A *signal generator* is a test device which generates a real sinusoid in the form of an electrical signal.[15] The output signal from a signal generator is specified in Hz, KHz, MHz or GHz. The phase usually cannot be specified; this makes sense since the signal generator and the device under test usually have no common time reference.

A signal generator can also be implemented in firmware or software (see the section on *NCO* at the end of this chapter). Such a generator provides samples rather than an electrical signal. In addition to a real sample stream, the NCO can generate a complex signal.

7.2 Complex Sinusoids

A real sinusoid can be modeled by painting a dot on a wheel, and then rotating the wheel and plotting the height of the dot as a function of time. The height will oscillate up or down. From the plot, we cannot determine the direction of rotation, but we can determine the rate of rotation.

A complex sinusoid has both real and imaginary components. These components are also called the *in-phase* and *quadrature* components (*I* and *Q*), or *IQ* components, where in-phase corresponds to the real component, and quadrature corresponds to the imaginary component.[16]

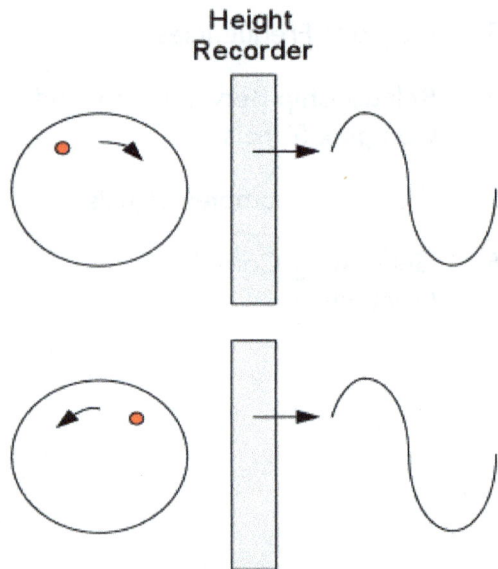

Figure 37. Direction cannot be determined from real signal plot.

_____

[15] Some signal generators are capable of generating complex outputs; these have two output ports, one for I, one for Q.

[16] The I and Q terms seem to have their origin in electronics engineering and phasor diagrams, whereas the terms "real" and "imaginary" have their origins in applied mathematics.

A complex sinusoid can be modeled by painting *two* dots on a wheel, each dot having a different color and being at 90 degrees from each other. The *I* and *Q* components are generated by rotating the wheel and plotting the height of the two different dots as a function of time.

Some signal generators *can* generate complex signals; these generators have two separate output connections.

The existence of real and complex sinusoids raises the question: *which should we work with? Ultimately, all signals are real, so why use complex signals?*

The answer is that many signal processing algorithms become easier to implement (and also easier to understand) when applied to complex signals. A complex signal is also known as an *analytic signal*.

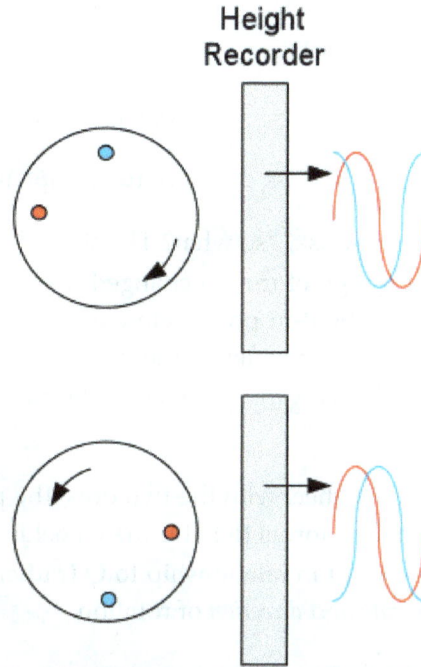

Figure 38. I and Q components allow determination of direction.

7.3 Negative Frequencies

Unless you've dealt with complex frequencies, the term "negative frequency" may seem unusual. If a real frequency is represented by cos (ωt), then replacing ω with $-\omega$ yields cos($-\omega t$), which is the same as cos(ωt), so there is no change whatsoever.

If the real signal is represented by sin (ωt), then replacing ω with $-\omega$ gives sin($-\omega t$), which is equal to $-$sin(ωt), which simply negates the signal. Based on this, there would seem to be no useful distinction between positive and negative frequencies.

We now consider the complex case, where the signal s is represented by $e^{j\omega t}$. Here we have:

$$s(t) = e^{j\omega t}$$

$$= \cos(\omega t) + j\sin(\omega t)$$

Negating the frequency gives:

$$s_{neg}(t) = e^{-j\omega t}$$
$$= \cos(-\omega t) + j\sin(-\omega t)$$
$$= \cos(\omega t) - j\sin(\omega t)$$

Right now one may ask, "so what? The sign of the *sin* component has changed." When the signal was real and the sign of the sin changed, we still had a sin wave, though inverted. In a complex signal, we have a built-in phase reference that travels with the sin wave: the cosine component. For a complex *positive* frequency, the sin reaches its peak value 90 degrees after the cosine reaches its peak value. For a *negative* frequency, the sin reaches its peak value 90 degrees *before* the cosine reaches its peak.

Going back to the wheel with the two dots, the plot of I and Q as a function of time now provide enough information for us to tell if the wheel is rotating clockwise or counter-clockwise. The fact that we can look at I in relationship to Q (rather than looking at a single sin(ωt) by itself) lets us determine the *rate* and *direction* of rotation.

Within the spectrum, a negative complex frequency is different from a positive complex frequency. In fact, a complex frequency of $-\omega$ shows up in a different place (to the left of 0 Hz) in the spectrum from the positive frequency ω. A spectrum analyzer will not show negative frequencies (since it deals with real signals), but an FFT performed on a set of samples will show this.

7.4 Relationship between Real and Complex Signals

A real signal is actually the *sum of two complex signals*. This is a characteristic of real signals: embodied within a real signal are two complex signals. As evidence of this, we consider the real signal cos(ωt):

$$\cos(\omega t) = (e^{j\omega t} + e^{-j\omega t})/2$$
$$= [\cos(\omega t) + j\sin(\omega t) + \cos(\omega t) - j\sin(\omega t)]/2$$
$$= (2\cos(\omega t))/2$$
$$= \cos(\omega t)$$

Therefore, cos(ωt) is the average of two complex frequencies, e^{jwt} and e^{-jwt}. [17]

---

[17] The astute reader may recognize this as the hyperbolic cosine of $j\omega t$.

96

A similar relationship can be shown for the real signal sin(ωt): this is the equivalent of

$$\sin(\omega t) = (\quad e^{j\omega t} - e^{-j\omega t})/(2j)$$

These relationships explain why the complex FFT of a real sinusoid of frequency ω shows peaks at both −ω and ω.

Amches Pointer

A real sample stream at a sample rate of 8K samples/sec can represent frequencies from 0 to 4 KHz; it provides a bandwidth of 4KHz. A complex sample stream at a sample rate of 8K complex samples per second can represent frequencies of -4KHz to +4KHz. The complex sample stream provides a bandwidth of 8 KHz.

7.5 Obtaining Complex Signals

Most signals originate from a single ADC that produces a stream of real samples at a fixed sample rate F_s. Many signal processing algorithms are simplified if we can represent the signal as a complex signal rather than the sum of two complex signals whose imaginary components cancel each other out.

As was mentioned above, a complex sinusoid gives us a sense of the frequency and also the "direction of rotation." A complex signal also provides unambiguous phase information: some of the energy will be present within the real component; some will be present within the imaginary component.

A complex sinusoid also gives an accurate magnitude measurement. Whereas a real sinusoid is characterized by undulations in magnitude, a complex sinusoid has a consistent magnitude from sample to sample. This characteristic simplifies, for example, AM demodulation, automatic gain control, and other signal processing operations.

Amches Pointer

A real signal **cannot** be converted to a complex signal by providing all zeroes for the imaginary component. Attempting to do so results in the sum of two complex signals, one having the negative frequencies of the other, which is typically not what is desired.

Complex Sampling

One way to get a complex sample stream is to perform complex sampling. This requires two ADCs, and a sample clock, with a phase-delayed version of the clock. If, for example, the sample clock operates at 100 MHz (10 nSec per sample), the phase delayed version of the sample clock is a 100 MHz clock having a 2.5 nSec delay from the in-phase version of the clock. This corresponds to a 90 degree phase shift of the sample clock.

The outputs from the two ADCs form the complex sample stream. In practice, this method is not often used: high sample rates make it difficult to generate an accurate phase shifted version of the clock, more hardware is required, and the results are susceptible to slight variations in the ADCs and the accuracy of the phase shifter.

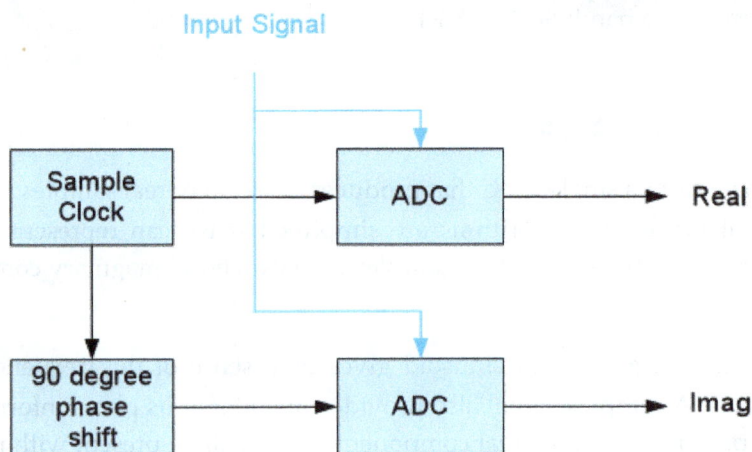

Figure 39. Complex Input Sampling.

Hilbert Transform Method

It is possible to generate a 90 degree phase shift for *each* frequency in the input signal, thereby creating the imaginary component. This phase shift is not a "simple delay". It is a frequency-dependent delay: low frequencies must be delayed 90 degrees, high frequencies must also be delayed 90 degrees, but a 90 degree delay for a low frequency is equal to a longer period of time than a 90 degree delay for a high frequency signal.

A special filter, called a Hilbert Transform, accomplishes this delay. In addition to the 90 degree phase shift (i.e., a frequency dependent delay), the Hilbert Transform introduces a group delay. Before pairing up the real signal with the phase shifted version, the real signal must undergo a group delay so that it aligns with the corresponding imaginary components.

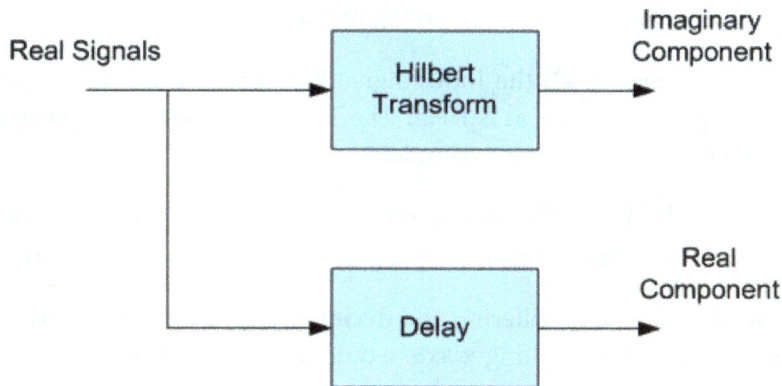

Figure 40. Hilbert Transformer converting Real to Complex.

Complex Mixing Method

We had mentioned above that a real signal is the sum of two complex signals: one having a positive frequency, the other having a negative frequency. When added together, the imaginary components cancel each other out, leaving the real signal.

The complex mixing method multiplies the real input signal stream by a complex sinusoid. This also downconverts (or upconverts) the signal, circularly shifting the spectrum. Shifting by ¼ of the sample rate, and then applying a low-pass filter having half of the spectral bandwidth[18] suppresses the negative frequency "images."

Shifting by ¼ the total spectrum is accomplished by multiplying the real signal by the complex frequency of $(-F_s/4)$. This complex frequency is the sequence (1,0), (0,-1), (-1, 0), (0, 1)… Since all samples are either $+/-1$ or $+/-i$, no multipliers are required.

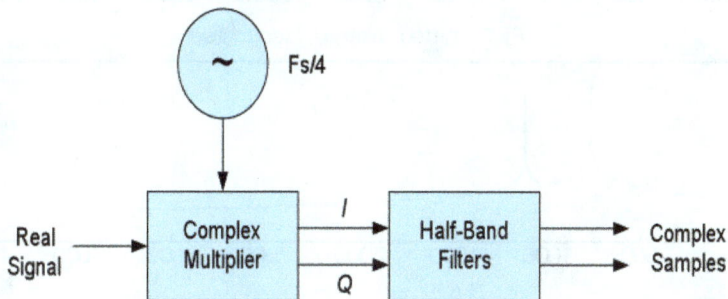

Figure 41. Complex Downconversion from Real to Complex.

---

[18] A half-band filter is a special FIR designed to do this efficiently

The filters are normally FIR low-pass filters having real taps.

This is illustrated in the Figure 42: the input signal consists of a 770 Hz signal plus a lower-amplitude 1120 Hz signal, sampled at 8 KHz. The spectrum shows the positive and negative frequency components, both of which are complex.

The spectrum is rotated by Fs/4, the middle plot shows the original frequencies on the X axis after rotation. The dotted red line indicates the frequency response of an ideal low-pass filter.

The final plot shows the result after filtering and decimation by two. Though the figure shows the x axis spanning 0 to 4 KHz, the resulting x-axis would span from -2KHz to 2KHz. If desired, the spectrum can be easily rotated again to place the frequencies in the positive region of the spectrum.

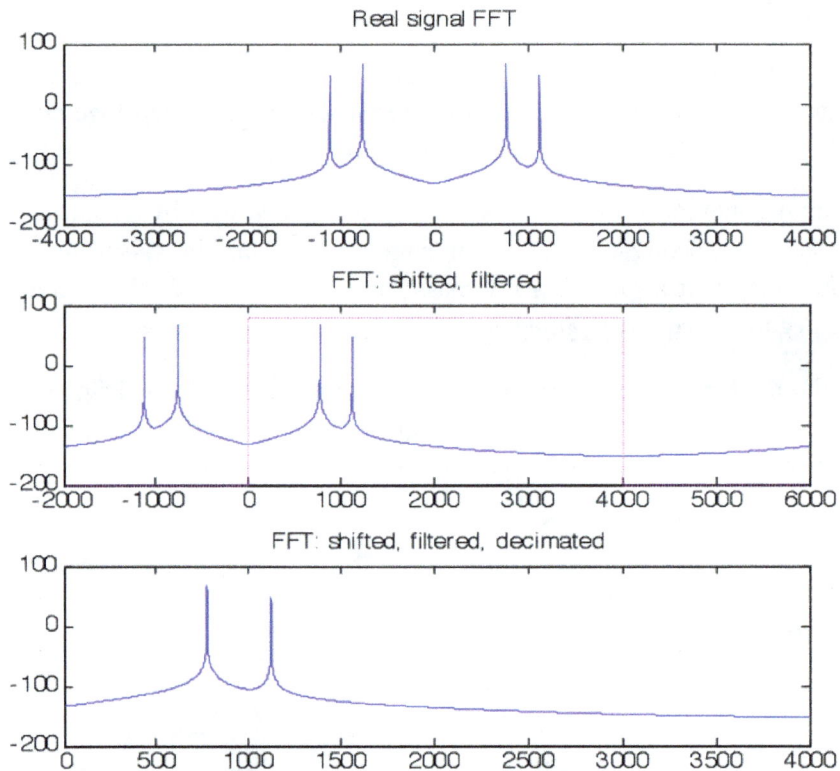

Figure 42. Conversion of a real signal to a complex signal.

7.6 Generating Complex Frequencies

A computational block that generates complex sinusoids having an arbitrary frequency is called a *Numerically Controlled Oscillator* (NCO).

We have already seen how a complex oscillator can be constructed by walking around the unit circle using successive complex multiplies. This method works, except that because the complex values are approximations, the accumulated error eventually forces the values off of the unit circle, where the errors continue to grow at an exponential rate.

A practical method of generating a complex sinusoid is by table lookup. Here, a memory contains a digitized single cycle of a complex sinusoid. For convenience the table length is a power of two, such as 128 samples. The NCO walks through the table, looking up an output value, and generating an output sample. Because the table contains a single cycle, the index can be allowed to wrap around to the beginning when it exceeds the table length. And because the complex values are confined to table entries, the results never get far off of the unit circle.

The output frequency is a function of the sample rate, the table size, and the index increment. For example, if the table contains 128 entries, and the index increment is one, the NCO will generate a full cycle for every 128 samples. The output frequency in this case is Fs/128. But if the index increment is 17, the NCO walks through the table 17 times faster, generating a frequency of Fs*17/127.

Figure 43. Complex lookup table of length 128 points.

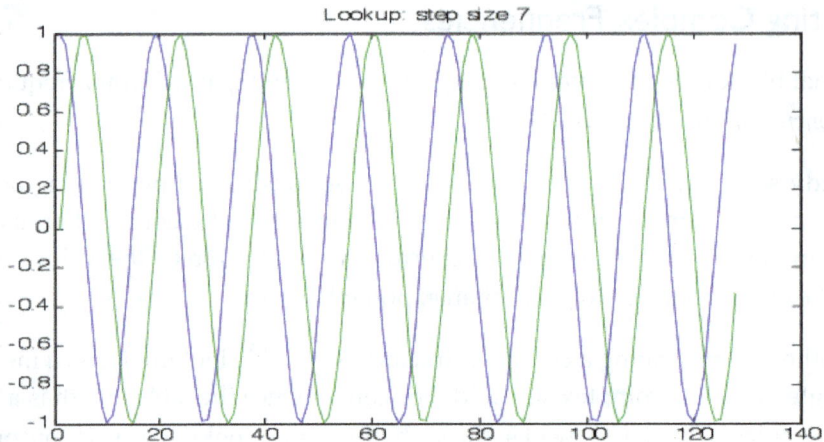

Figure 44. Cos/Sin lookup, step size 7, table size 128.

The maximum frequency occurs when the table increment is (table size)/2, or 64 for our example. The maximum frequency that can be synthesized is Fs/2. To generate a signal having frequency f, use the following formula to determine the index increment:

increment = (Table Size)/(Fs/f).

Example:

> A lookup table has 1024 entries. The sample rate is 10 MHz. What index increment is needed in order to generate a sinusoid having a frequency of 625 KHz?
>
> Answer: increment = (1024)/(10/0.625) = 1024/16 = 64. The frequency 625 KHz is 1/16th of the sample rate. At this sample rate, the NCO must generate a full cycle for every 16 lookup values, so the index increment is 1024/16 = 64.

Table entries are normally complex values. Also, there is no reason that the table increment need be positive: it can be negative. This results in the generation of a negative frequency, which is useful for down converting a signal.

NCO Accuracy

There are three factors that limit the NCO accuracy:

1. Table Size: larger tables allow finer-grain lookups;

2. Table Width: the bit width of each entry should be sufficiently large;

3. The index increment should be an integer.

102

Quite often these factors are difficult to accommodate. One or more of the methods below can then be applied to improve accuracy.

Larger Tables

Because the lookup table must be a power of two in length, increasing the size of the table means doubling (or quadrupling, or more) the size. This allows a finer-grain lookup. It also increases the number of integer index increment values that are available.

This is nice if an FPGA supports it, but some FPGA designs are memory constrained. Since the table size wants to be a power of two, the next largest table requires twice the storage of the existing table. This is not always possible. Alas, not all is lost. The techniques below can compensate for shorter tables.

Table Bit Width Entries

When the index increment works out to be an integer, the *only* limiting factor for the output accuracy of the NCO is the bit width of each table entry. Here, we can assume that each bit of width provides 6 dB of accuracy, so a table of 16 bit wide entries provides 96 dB of accuracy.

By the way, there is an easy way to get an extra bit with sin/cos lookup tables. Consider 16 bit entries. The values 1.0 and -1.0 occur at exactly four places within the table. These values are the "widest" in the table. When they occur, low bits are not used. This means that the table width has to provide a single extra bit merely to accommodate these values. If the lookup values are first scaled by, say, 0.999, then all values in the lookup table span the range -0.999 to 0.999. This makes better use of the available entries, and purchases almost 6dB more of table accuracy.[19] Users of an NCO so designed may notice that the NCO has a loss of 0.008 dB compared with the ideal output values; users so concerned can usually compensate for this in a later processing step, if desired.

When the index increment is not an integer, the lookup wants to be "between" entries, and choosing the nearest entry degrades the accuracy of the result.

Non-Integer Increments

Integer increments are desirable, but an NCO is often called upon to generate frequencies that translate to non-integer increments.

The approach is straightforward: we design the index increment (which we now call the phase increment) to be a fixed-point value, having $\log_2$(table size) bits to the left of the radix point, and several bits to the right of the radix point. The phase increment can now accommodate a non-

---

[19] *Recovers* may be a better word than *purchases* here: this accuracy was lost when the table was populated with the full range of sin/cos values, including +1 and -1.

integer value. Every time the NCO generates an output sample, it adds the phase increment to the phase accumulator, which has the same alignment as the phase increment. The most significant $\log_2$(table size) bits of the phase accumulator become the index into the table.

This method works well, and it is inexpensive, since the only "wide" values in the implementation are the phase increment and phase accumulator. When the phase increment has a nonzero fractional part, the fractional part will build up in the accumulator, until it overflows into the integer part, nudging the table index up every now and then.

Non-Integer Increments > 1

When the increment is an integer plus a nonzero fraction, many index values will be "between" table entries. The error is approximately the fractional index value times the slope of the lookup function at the integer point:

$$\text{error} \approx \text{frac(index)} \cdot \frac{d}{dx}\, e^{2\pi j x / \text{Tsize}}$$

The maximum error for cosine will occur near $x = 0$ and $x = \text{Tsize}/2$, and will be as large as $2\pi/\text{Tsize}$. The maximum error for the sin term will occur near $x = \text{Tsize}/4$ and $x = 3\text{Tsize}/4$, and will be as large as $2\pi/\text{Tsize}$. These error figures can be halved if the table lookup rounds the index to the nearest value.

The error approximation sets a limit for the accuracy of the table entries. For a table of size T, table entries should be at least $\log_2(2\pi/\text{Tsize})$ bits. Unless some form of interpolation is to be used, there is little to be gained by having more than one or two bits beyond this.

Non-Integer Increments < 1

A fractional increment generates an output signal having a frequency lower than F_s/Tsize. In this case, the table index will remain at x for several samples, until n times the accumulated fractional value results in a new integer index.

Figure 45 shows a lookup with a step size of 0.5. Some "graininess" (quantization) can be seen in the plot. This is because a step size of 0.5 results in an index increment for every two output samples generated.

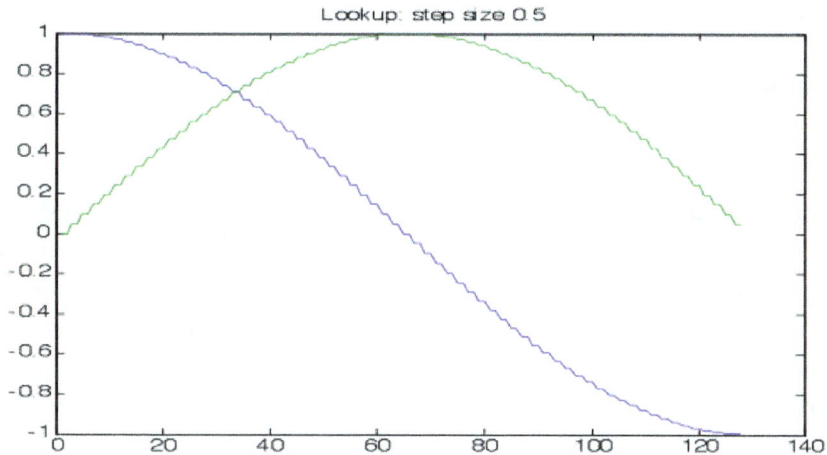

Figure 45. Cos/Sin lookup, table size 128, step size 0.5.

Phase Jitter

Fractional extension of the phase increment results in a near-perfect NCO output frequency[20]. However, the accumulation of fractional bits and the "nudging" occur at regular intervals. The result is an "adjustment" where the output sinusoid periodically "jumps" based on the accumulated fractional bits. Because this happens at regular intervals, the resulting spectrum shows "spurs."

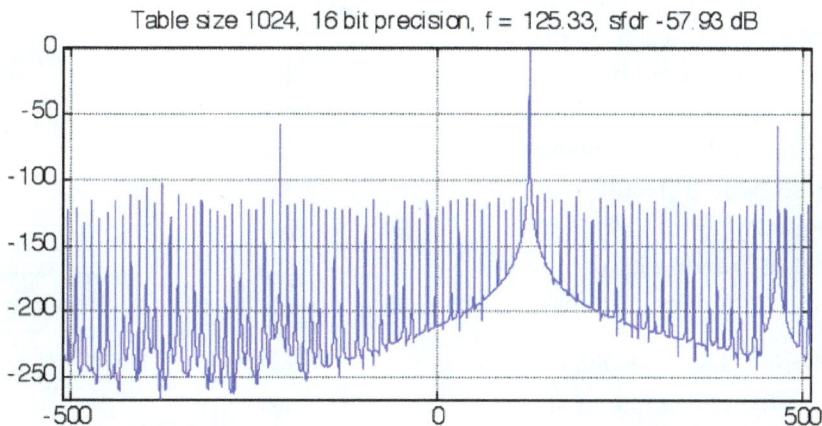

Figure 46. Non-integer phase increment and resulting spurs.

---

[20] That is, if frequency is measured only by the number of zero-crossings per unit time, and the spectral quality of the result is otherwise ignored.

The figure above shows multiple spurs, the highest of which are 58 dB below the desired signal. The power difference between the desired output peak and the next highest spur is the *spur-free dynamic range* (SFDR). At 16 bits of table precision, we expected about 96 dB of performance.

Why care about spurs?

Often, the output of an NCO will be mixed with an input signal to center-tune the signal down to 0 Hz. If the NCO contains a single, clean frequency, this will mix with the input signal and bring the desired signal down to 0. For each spur in the NCO output, another signal, from another location within the spectrum, will also be mixed down to 0 Hz. Now, if the signal that we desire to mix down is weak, and the signals mixed down by the spurs are strong, the total noise centered at 0 in the resulting signal will be significant. A clean NCO is therefore *highly* desirable, if not required.

Improving Performance

An interesting and inexpensive way to improve performance is to incorporate *phase dithering* within the NCO. Recall that the problem is a non-integer phase increment. Recall also that the fractional phase increments accumulate until they "kick over" to the next integer value. This happens at regular intervals.

Phase dithering is a process whereby the NCO adds a pseudo-random number to the fractional part of the accumulated phase for each entry before performing the lookup. It is critical to note that the random value is not accumulated with the phase, though (doing so would result in a cumulative frequency error). Instead, this has the effect of "randomizing" the times at which the "kick over" occurs.

The phase dithering has two effects: first, it breaks up the coherency of the fractional overflow, reducing the amplitude of any spurs. Second, it raises the noise floor elsewhere. It essentially adds low level noise, while spreading out the energy in the spurs.

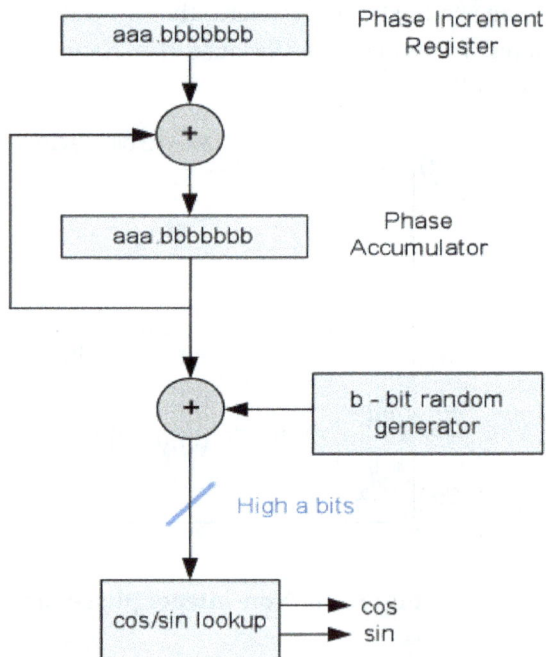

Figure 47. An NCO with Phase Dithering.

The result of dithering is seen in Figure 48. The spurs, which were at -58 dB, have moved to -85 dB; an improvement of 27 dB using the same table size and precision.

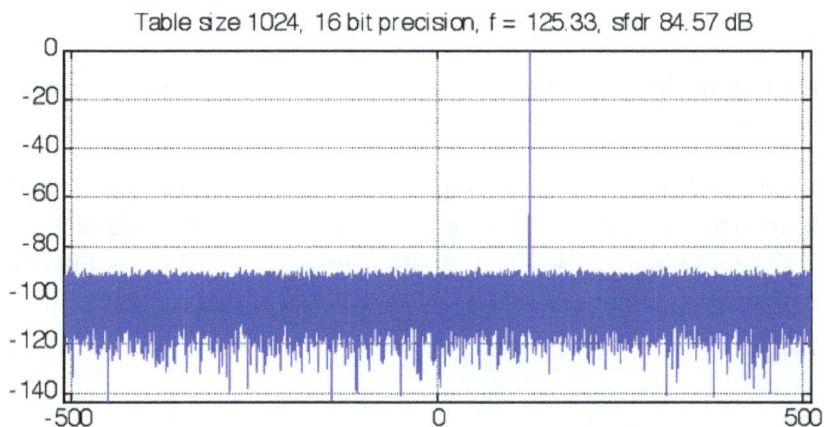

Figure 48. NCO performance with phase dithering.

Other methods exist for improving performance, and also for reducing the table size. For an excellent description of a specific NCO implementation, see Graychip[2].

Exercises

1. Describe a plot of the magnitude of the FFT of the point-by-point product of $e^{2\pi j 10t}$ and $\cos(2\pi t)$. How many frequencies are present?

2. Describe a plot of the magnitude of the FFT of the point-by-point product of $e^{2\pi j 10t}$ and $e^{2\pi j t}$. How many frequencies are present?

3. A poorly designed NCO produces frequency -f1 at 0 dB, and a spur at frequency -f2 and level -30 dB. This is mixed with a signal having a level of 12 dB at f1, and a stronger signal at f2 having a signal level of 28 dB. The result is low-pass filtered using a filter having 0 dB of attenuation around 0 Hz. Describe the signal content of the result.

References

[1] Lyons, Richard G., *Understanding Digital Signal Processing*, Second Edition, Prentice Hall, Upper Saddle River, New Jersey, 2004, pp. 200-207.

[2] Graychip *GC4016 Multi-Standard Quad DDC Chip Data Sheet*, Revision 1.0, Texas Instruments, 2001, pp. 6-7.

The FFT

8

In this chapter...

A common problem in signal processing is to determine what frequencies are present in a signal.

The Fourier Transform translates a signal from a set of samples occurring at a fixed sample rate into the equivalent spectrum (that is, the frequency components). The spectral components, when recombined, form the original signal.

A developer can often use an FFT library function, or an FPGA block, that computes the FFT. Knowing the inner workings of the FFT is necessary in order to control the resolution, precision, and resources required to compute the FFT of a signal.

8.1 Equivalence between Time and Frequency Domain

Any signal can be represented as the sum of sines and cosines of different frequencies, phases and amplitudes. Furthermore, these sines and cosines of different frequencies, phases, and amplitudes can be added together to recreate the original signal.

A plot of a signal as a function of time, as one would observe on an oscilloscope, is said to be a *time domain* signal. The x-axis is expressed in units of time, the y-axis is amplitude.

Since a signal can be represented as the sum of sines and cosines having different amplitudes and phases, the question becomes, how do we find out what the amplitudes and phases are for each different frequency? A plot of the amplitudes as a function of frequency, as one would see on a spectrum analyzer, is said to represent the *frequency domain*. The FFT is the means by which we determine the amplitudes and phases for the different frequencies that form the original signal.

8.2 Uniqueness

While a signal *can* be represented as the sum of sines and cosines, a question becomes, for a given signal, is there a *unique set* of sines and cosines that, when added together, form the signal? The answer to this question is *yes*. This is important: it means that for any signal, its Fourier transform is unique.

It also implies that we can use the inverse transform to get back to the original time domain sequence. Once we know the amplitudes and phases, we can reconstruct the exact input signal by adding sines and cosines having the specified amplitudes and phases.[21] The time and frequency domain representations of a signal are therefore equivalent, in the sense that no information is gained or lost when moving from one domain to the other. (We do, however, gain insight into the frequency components by examining the frequency domain, but we do not change the information content.).

A transform is exactly that: it transforms the data from one form to another[22]. No information is lost in the process; nor is information gained.

8.3 The FFT and the DFT

Normally the *fast Fourier transform* (FFT) is employed to convert time samples to frequencies. The Discrete Fourier Transform (DFT) may also be used. The FFT is a performance enhanced version

---

[21] Except for round-off errors that occur within the computation.

[22] There are other transforms: Laplace, Walsh, Wavelet, Hadamard, z, etc. Fourier is certainly the most popular within signal processing.

of the Discrete Fourier Transform (DFT). Both give the same results (apart from rounding errors); but the FFT offers a huge performance advantage.

The distinction between the DFT and the FFT is similar to the distinction between the quicksort algorithm and an exchange sort algorithm: both algorithms give the same answer. One is easier to code. The one that is harder to code is much faster.

An FFT, unlike the DFT, may be limited to vectors having specific lengths, depending on the implementation. For example, a radix-2 FFT (the most common type) operates on vectors having length of 2^n.

How It Works

To understand how the Discrete Fourier Transform works, we refresh our memory regarding orthogonal vectors: two vectors are orthogonal if their dot product is zero. We also note that the dot product of a vector with itself is nonzero (unless the vector is all zeroes). We then recall that a vector consisting of an integer number of cycles of a complex sinusoid has the following properties:

- The dot product of the vector with another vector having the same length but a different number of cycles is zero;
- The dot product of the vector with itself is nonzero.

Finally, we recall that when computing the dot product of two complex vectors, we use the conjugate of the second vector.

We then assume that the time domain signal can be represented as the sum of several complex sinusoids of different frequencies 0, ω_1, ω_2, etc:

$$S_t = A_0 + A_1 \cdot e^{j\omega 1t} + A_2 \cdot e^{j\omega 2t} + \ldots$$

The A coefficients are complex values that we seek to find. A_0 is the DC value. A_n is a complex number representing the magnitude and phase of frequency ω_n that is present within S_t. If the signal contains no energy at frequency ω_n, then A_n will be zero.

We now find A_n for different values of n. We use the fact that a complex sinusoid of frequency ω_n is orthogonal to one of frequency ω_m when n ≠ m as a "filter" to isolate A_n. We compute the dot product of both sides with $e^{j\omega nt}$

$$S \cdot e^{j\omega nt} = [A_0 + A_1 \cdot e^{j\omega 1t} + A_2 \cdot e^{j\omega 2t} + \ldots] \cdot e^{j\omega nt}$$

Due to the orthogonal relationship between complex sinusoids, *all terms on the right vanish to 0 except the term $A_n \cdot e^{j\omega nt}$:*

$$S \cdot e^{j\omega nt} = A_n \cdot e^{j\omega nt} \cdot e^{j\omega nt}$$

We then recall that when computing dot products of complex vectors, we use the conjugate of the second term. For a complex exponential, the conjugate is the same as negating the exponent:

$$S \cdot e^{-j\omega nt} = A_n \cdot (e^{j\omega nt} \cdot e^{-j\omega nt})$$

Since $e^x \cdot e^{-x} = 1$, the exponent on the right vanishes, leaving us with the formula for finding A_n :

$$A_n = S_t \cdot e^{-j\omega nt}$$

We repeat this for all frequency values of n to get the entire spectrum. The term A_n is essentially the correlation between a complex sinusoid of frequency n, and the input signal.

8.4 The Goertzel Algorithm

We diverge here for a moment to discuss the *Goertzel Algorithm*. The Goertzel algorithm is the dot product between an input signal and a complex sinusoid of a single fixed frequency.[23] The resulting output is a single complex value indicating the amplitude and phase of the frequency within the input signal.

The Goertzel algorithm is of interest when we wish to determine the amplitude of a small number of frequencies within an input signal: if the number of frequencies N_f is small, then performing N Goertzel algorithms often requires fewer operations then performing an FFT and examining N peaks of the result, unless N is "large."

The Goertzel algorithm for frequency $\omega = 2\pi f / F_s$ and an N point vector is a single complex value computed by:

$$G_\omega = \sum_{k=0}^{N-1} (x_k)(e^{-j\omega k})$$

The summation is recognizable as the dot product of the input sequence and a complex sinusoid of frequency ω. The multiplier outside the summation has magnitude one; it changes the phase of the result. If the number of points N spans an integer number of cycles at frequency ω, or if the phase of the result is of no interest, then the phase multiplier can be ignored.

---

[23] Be sure to use the negative frequency for the dot product.

Recursive Form

The complex sinusoid may be a pre-computed table, or, if the vector length is "small," the complex sinusoid may be computed "on the fly" by walking around the unit circle in the complex plane (see 72). Alternately, the transform can be computed using the recursion:

$$y_{-1} = 0$$
$$y_n = (y_{n-1})(e^{j\omega}) + x_n, \qquad \text{for } n = 0..N-1$$
$$G_\omega = (y_{N-1})(e^{j\omega})$$

Expanding the recursion gives:

$$G_\omega = e^{j\omega} ((((x_0 e^{j\omega} + x_1) e^{j\omega} + x_2) e^{j\omega} + x_3) e^{j\omega} ... + x_n)$$

Collecting complex multipliers then gives:

$$G_\omega = e^{j\omega} (x_0 e^{j\omega(n-1)} + x_1 e^{j\omega(n-2)} + x_2 e^{j\omega(n-3)} + ...)$$
$$= (x_0 e^{j\omega n} + x_1 e^{j\omega(n-1)} + x_2 e^{j\omega(n-2)} + ...)$$

When n/N is an integer, then the complex sinusoid spans an exact number of cycles, and $e^{j\omega n}$ becomes one, so:

$$G_\omega = (x_0 + x_1 e^{j\omega(-1)} + x_2 e^{j\omega(-2)} + ...)$$
$$= \sum_{k=0}^{N-1} (x_k)(e^{-j\omega k})$$

Therefore, rather than computing a series of complex coefficients, the Goertzel algorithm can be computed using a single complex multiplier ($e^{j\omega}$). The partial sum is rotated by the complex multiplier before adding each new sample of the input.

Scaling by Length

The formula for the Goertzel algorithm at frequency f of a vector x sampled at F_s:

$$G_f = (1/N) \sum_{k=0}^{N-1} (x_n e^{-2\pi jk(f/Fs)})$$

Here we have assumed that the vector length contains an integer number of cycles of frequency f/Fs, and we have divided by the length to obtain the per-sample amplitude.

Tuning the Goertzel Algorithm

The vector length of the Goertzel should be "tuned" to so that the vector length can contain an integer number of cycles at the frequency of interest. Otherwise, the transform result will not peak exactly at the desired frequency. If an exact integer multiple is not attainable, then the number of samples that comes close to the zero crossing for an integer number of cycles should be used.

Example: At 8K samples per second, a single cycle of 440 Hz occupies 18.1818 samples. At this sample rate, 100 mSec requires 800 samples of data. If we wanted to examine a 100 mSec set of samples to determine the 440 Hz content, the Goertzel algorithm with a length of 800 samples will span exactly 44 cycles.

If, however, we wish to determine the 852 Hz content of a 100 mSec segment, each cycle now occupies 9.39 samples. 800 samples contain 85.2 cycles. Rather than use 800 samples, we can use 798 samples, which contain 84.987 cycles. While we are examining slightly less than 100 mSec, the correlation against 852 Hz will be more precise.

Percent Power

Parseval's Relation states that the time domain power is equal to the sum of the power in all frequency components (see Kaplan[1]). The result of the Goertzel algorithm can be translated allowing us to determine what percent of the input signal is represented by frequency component G_f:

$$P_{avg} = (1/len) \sum [x_n \cdot conj(x_n)]$$

$$G_{norm} = [G_f \cdot conj(G_f)] / P_{avg}$$

Of course the length used to compute power must be the same length used to compute the Goertzel algorithm. The result will be between 1.0 and 0. If the input signal is real, the result will be between 0.5 and 0.

Goertzel Selectivity

Any length can be used for the transform, with more precise results occurring when the length is approximately an integer number of cycles of *f* at the sample rate. A longer transform results in a narrower response; a shorter transform results in a wider response. Figure 49 shows the magnitude of the Goertzel algorithm centered at 693 Hz for transform lengths of 312, 265, 208, and 161 samples, at a sample rate of 8K samples/second. The first plot shows the linear response, the second shows the response in dB. The narrower curve is a result of the longest transform length. The 3 dB points vary from 5 Hz wide for the shortest transform to 2 dB wide for the longest transform.

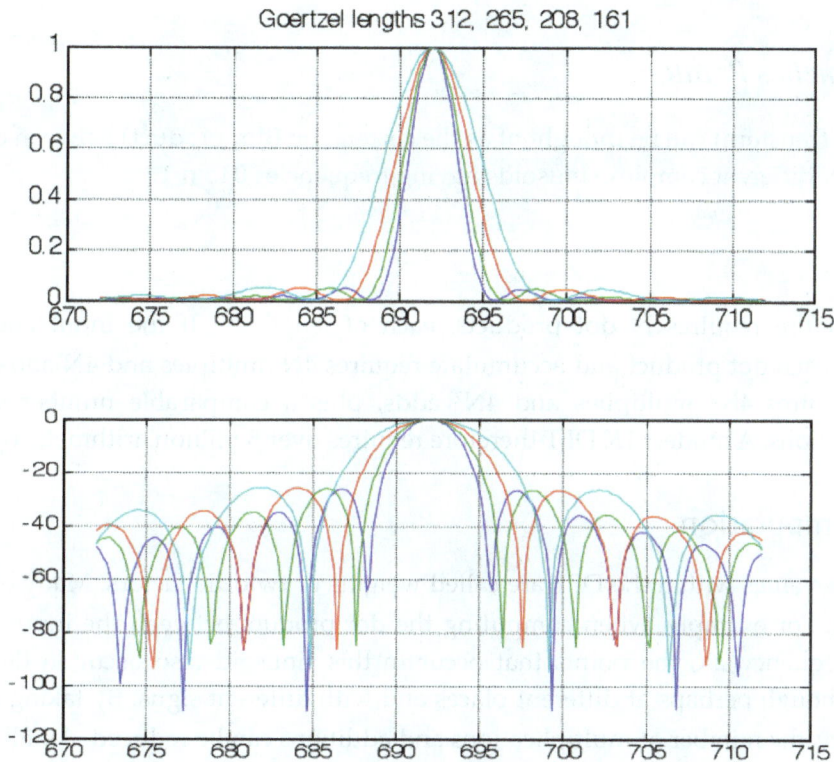

Figure 49. Goertzel Selectivity as a function of Transform Length.

115

8.5 DFT Computation

For N input points, the DFT produces N output points, called "bins." Each bin is a complex number:

- The magnitude of the complex value in bin K represents how much energy at frequency K was present in the input signal;

- The phase of the complex value in bin K represents the phase, relative to the start of the sample set, of the frequency component K within the signal.

Output K is computed by computing the dot product between the input data, and a complex sinusoid of length N points that has K complete cycles spread over the N points[24]. This is repeated for K different frequencies, producing K complex output points. Said differently, the DFT of a signal is computed by applying the Goertzel algorithm N times for frequencies 0 to N-1.

> ### *Amches Pointer*
>
> The Fourier Transform can be thought of as the correlation (dot product) between the input signal, and N different complex sinusoids having frequencies 0 to n-1.

The DFT therefore requires N dot products, each of length N. If the input and output are complex, then each dot product and accumulate requires 4N multiples and 4N adds. Computing all N bins requires $4N^2$ multiplies and $4N^2$ adds, plus a comparable number of loads and indexing operations. A modest 1K DFT therefore requires over 8 million arithmetic operations.

8.6 FFT Computation

The *sin* and *cos* values within the DFT are called weights or twiddle factors. Many of the weights are redundant. For example, when computing the dot product between the input vector and a sinusoid of frequency 2X, the points that occur in this sinusoid also occur in the sinusoid of frequency X, though perhaps at different places and with different signs. By taking advantage of the redundancy, the number of multiplications and additions can be reduced. An FFT does this: it reduces the number of computations required to compute a DFT by exploiting redundancy in the complex sinusoids.

---

[24] The complex sinusoid must have a negative sin component.

The largest amount of redundancy occurs when the number of samples (N) is a power of two. Therefore, most FFT implementations are designed for operation on vectors of length $N = 2^n$. If the number of samples is a prime number, then there is very little redundancy to eliminate. For composite N, algorithms exist that are faster than the DFT, but slower than those for length 2^n.

For an even length, the FFT can be computed recursively by computing the FFT of the even points, and the odd points. The results are multiplied by a weighting factor, and then combined (added/subtracted) to form the final results. The recursive decomposition into even and odd points is performed n times for a length of 2^n, resulting in n stages, each of which processes of 2^n points. For details, see Oppenhiem[2].

For length 2^n, the number of multiplies and adds reduces from $8N^2$ to approximately $5N \cdot \log_2(N)$. For 1K complex samples, the FFT requires about 50,000 operations, while the DFT requires 8 million operations. The FFT is therefore 160 times faster for a 1K vector. As the number of points increases, the performance improvement factor also increases: a 4K FFT versus DFT requires 246K versus 128 million operations; the FFT is 325 times faster in this case.

Amches Pointer

The FFT and DFT results for a given vector will match, except for slight differences due to rounding errors. The FFT will generally be more accurate, and faster. The FFT is generally harder to code, and requires that either the results or inputs be re-ordered.

For software FFT implementations, there is a sharp decline in performance when the size of the complex vectors and size of the weights exceed the size of data caches within the processor. This is true for both DFT and FFT implementations.

A number of excellent resources detailing the FFT algorithm implementation and its variants exist. Given that most engineers will be *users* of the FFT rather than implementers, we refer interested parties to these resources.

A consequence of exploiting the symmetry is that the inputs to the FFT (or outputs from the FFT) must be re-ordered in order to accommodate the algorithm. For a radix-2 FFT, the inputs (or outputs) must be re-ordered using *bit-reverse* indexing. Most software library functions perform the necessary re-ordering transparently; most FPGA functional blocks generate re-ordering as an option.

8.7 FFT, DFT Bin Ordering

Bin 0 of an FFT/DFT is always the DC component. Bins 1 to n/2 represent positive frequencies. Bin n/2 is the Nyquist frequency; it contains the energy for the highest frequency component realizable at the given sample rate.

Bin n-1 is the last bin in the result. This corresponds to the first negative frequency; n-2 is the second negative frequency, etc., all the way down to bin n/2.

Bin Numbering for Real Signals

When the input signal is real, FFT bins 1 to n/2 are the mirror image of bins n/2 to n-1. This is because each component having real frequency f is the sum of two complex frequency components having frequencies f and –f. Therefore, if the input signal is real, it is customary to plot only n/2 bins of the result.

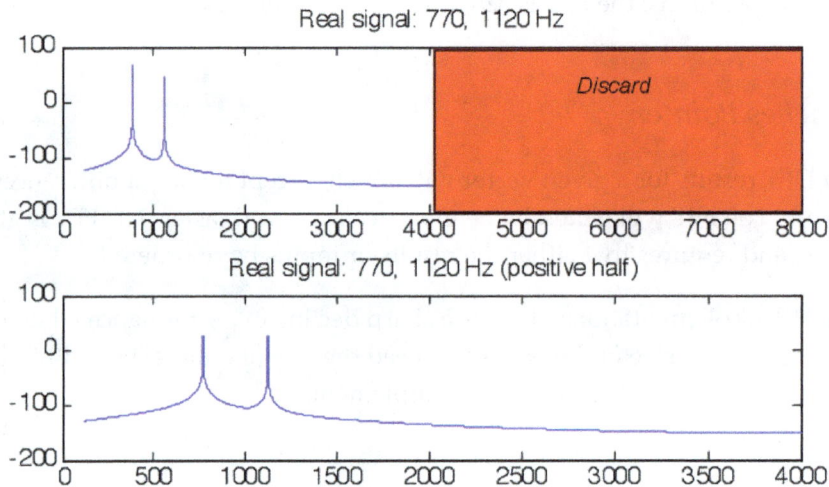

Figure 50. The second half of a real FFT is redundant, and often not plotted.

Bin Numbering for Complex Signals

When the input signal is complex, it can contain both positive and negative frequencies. In this case, the first n/2 bins are *not* the mirror image of the second n/2 bins. For a complex signal, it is customary to rotate the bins by n/2 points before plotting a display. This places the 0 Hz (DC component) in the center of the plot, with negative frequencies to the left, and positive frequencies to the right.

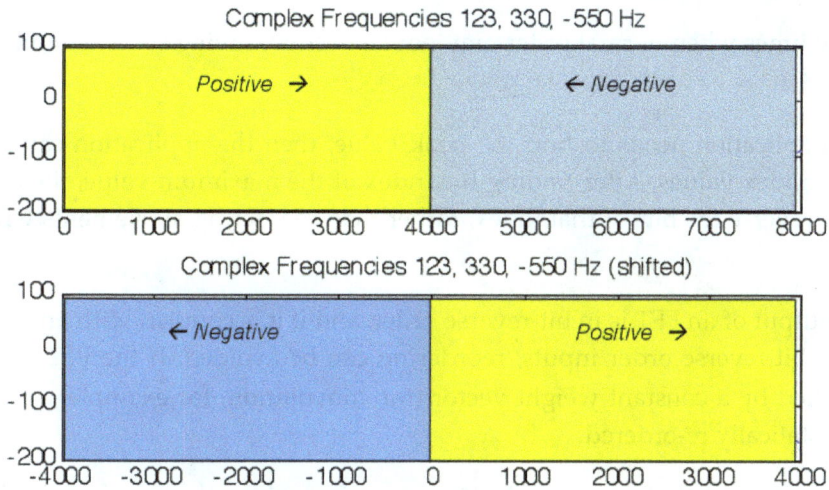

Figure 51. The complex FFT results are rotated, placing bin 0 in the center.

Bit-Reverse Index Order

The results of a radix-2 FFT (or the inputs, depending on the implementation), may be ordered in bit-reverse-index ordered. This is a byproduct of the algorithm. FPGA implementations may give the option of producing the output in bit-reverse order or natural order. Natural order requires slightly more logic, but presents output bins in the order 0..n-1.

Bit-reverse indexing is accomplished by reversing the bits in each possible index value. The table below shows the bit-reverse index values for a 3 bit index.

Input Index		Bit-Reverse Index	
Decimal	Binary	Binary	Decimal
0	000	000	0
1	001	100	4
2	010	010	2
3	011	110	6
4	100	001	1
5	101	101	5
6	110	011	3
7	111	111	7

There are a few times when normal order may not be needed, saving some amount of processing time and/or logic:

- If the application needs to find the peak value, then the application can search the bit-reverse-index values. After finding the index of the maximum value, the application can simply reverse the bits of that index, rather than reordering all elements within the input array.

- If the output of an FFT is in bit-reverse order, and if it is coupled with an inverse FFT that accepts bit-reverse order inputs, reordering can be avoided. If the FFT results are to be multiplied by a constant weight vector (for convolution, for example), then the weights can be statically re-ordered.

- In the case of coherent or non-coherent averaging, where the bins from multiple FFTs are added together to produce a single vector, the FFT results can remain in bit reverse order. Once all FFTs have been summed, the final summation vector can be converted to normal order if needed.

The cost of reordering to normal order is small; unless space and processing time are tightly constrained, reordering to normal order is recommended.

8.8 FFT, DFT Resolution

The FFT of a vector of length N decomposes a signal into N distinct complex sinusoids having frequencies of 0 (DC), 1, 2... N-1. For a vector of length N, a frequency component f contains exactly F cycles that span the N points within the vector.

Each of the N resulting bins represents a range of frequencies. If the input vector represents a signal sampled at sample rate F_s hertz, then each bin of the FFT has a width of F_s/N hertz. A wider FFT, operating on more samples, provides greater resolution.

Example: For a signal sampled at 8K samples per second, the resolution for a 512 point FFT is 15.625 hertz per bin; a 1K FFT will provide a resolution of 7.8125 hertz per bin.

The input signal is not constrained to having frequencies of 0, 1, 2, .. N-1 cycles. For example, an input signal may have 3.2 cycles spread over 256 points. A 256 point FFT would show a "widening" around bin 3: some energy will be in bins 4 and 2, less in bins 5, and 1, etc. An FFT of length 1280 (5 times 256) would contain exactly 16 cycles for the same signal, and would show a precise peak in bin 16, with zeroes in the adjacent bins.

8.9 FFT, DFT Gain

Many FFT implementations provide an output that is *length* times the expected result. This gain factor is equal to $6 \cdot \log_2$(length) decibels. This occurs because each output bin is the sum of *length* weighted entries. The gain is of no consequence if the user intends to examine the ratio of peak values with noise floor values; the peak and noise floor values will both be increased by $6 \cdot \log_2$(length) but the difference will be the same.

Some FPGA toolkits provide an option for intermediate scaling during FFT computation: results are shifted right one bit after each of the $\log_2$ stages, resulting in a 0-dB gain.

If the inputs to a complex FFT are fractional values, and if all fractional values fall on or within the unit circle, then the *magnitude* of the output for each bin will never exceed 1.0, apart from rounding errors.

8.10 Periodic Extension

The FFT (and DFT) are often applied to a "record" of samples extracted from some data stream. While the transform operates on this data, it assumes that the data extends and repeats infinitely to the left and right. This is called the *periodic extension* of the function, shown in the figure on the next page.

Amches Pointer

An FFT of length N can represent frequencies as low as f_s/N, not including a DC component. An FFT can represent frequencies no higher than $f_s/2$. Longer record lengths result in more output bins, and finer-grain resolution of bins.

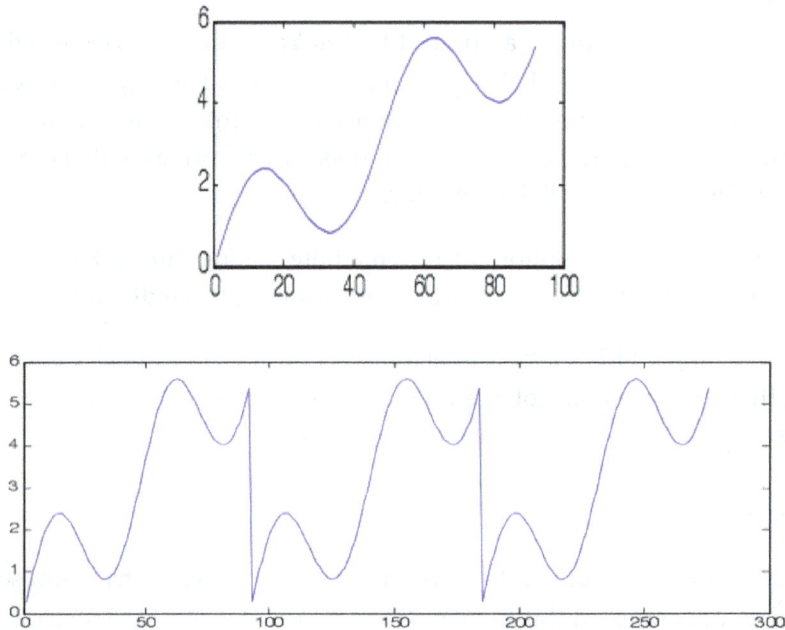

Figure 52. A signal and its periodic extension showing "jump discontinuities."

The periodic extension introduces "jump discontinuities" at the splice points. Though not in the original data set, these discontinuities introduce significant high frequency components within the FFT. The FFT is still correct; however, one must keep in mind that the FFT is that of the periodic extension of the original signal, not just the FFT of the limited data set.

Amches Pointer

Another way to view periodic extension: consider the fact that for an n-point FFT, there are n frequencies: a DC component, a fundamental, and n-2 harmonics of the fundamental. Since the signal is represented by the sum of n frequencies having different amplitudes and phases, and since they are harmonically related, the result will have a period of n points.

8.11 FFT, DFT Windowing

A common way to reduce the effects of discontinuities from periodic extension is to multiply the input signal by a *window function* before computing the FFT (Lyons[3]). A window function:

- Is normally a real, symmetric, smooth function that tapers off at the edges.

- Reduces the jump discontinuities introduced by periodic extension. This is desirable.

- "Widens" the spectral peaks that we would like to see. This is an undesirable but tolerable effect.

A window function, and its effect on the FFT of data, is shown below:

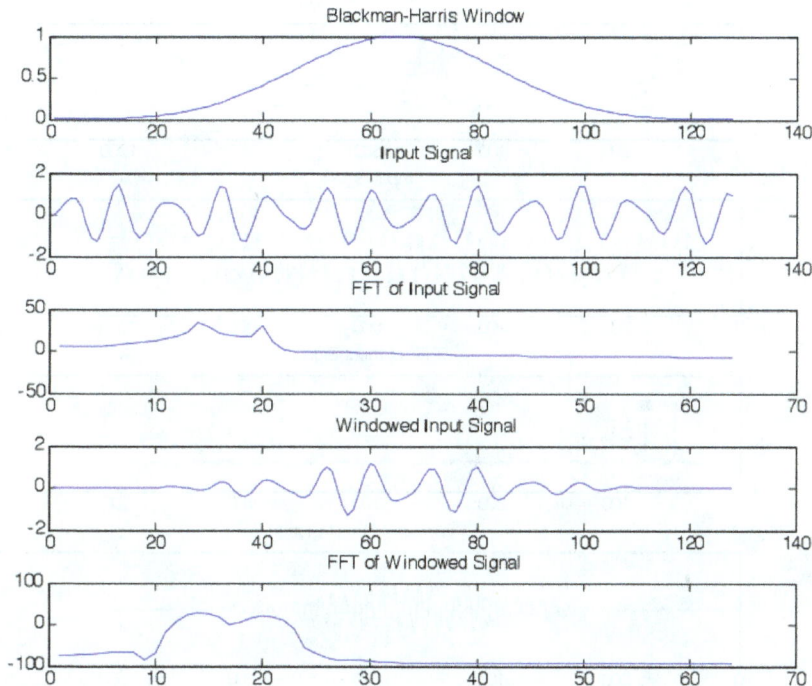

Figure 53. Windows improve peak level but widen the peaks.

Note the FFT results without windowing: the peaks are about 35 dB above the "noise", and there is significant energy across the spectrum. However, after windowing is applied, the peaks are more prominent in amplitude, but wide; the noise floor across the spectrum is reduced. High frequency components that were artificially introduced as result of the jump discontinuity are suppressed.

The above signal was generated using sinusoids having 13.3 cycles per 128 points and 19.17 cycles per 128 points. Neither of these signals fit "nicely" within 128 points, so jump discontinuities as a result of periodic extensions are certain.

The plots below show a different example. The signals have *exactly* 52 and 76 cycles. Because the signals fit *exactly* within the 1024 data points, the periodic extension has no jump discontinuities. The un-windowed FFT results are therefore sharp. The windowed results show widening of the peaks.

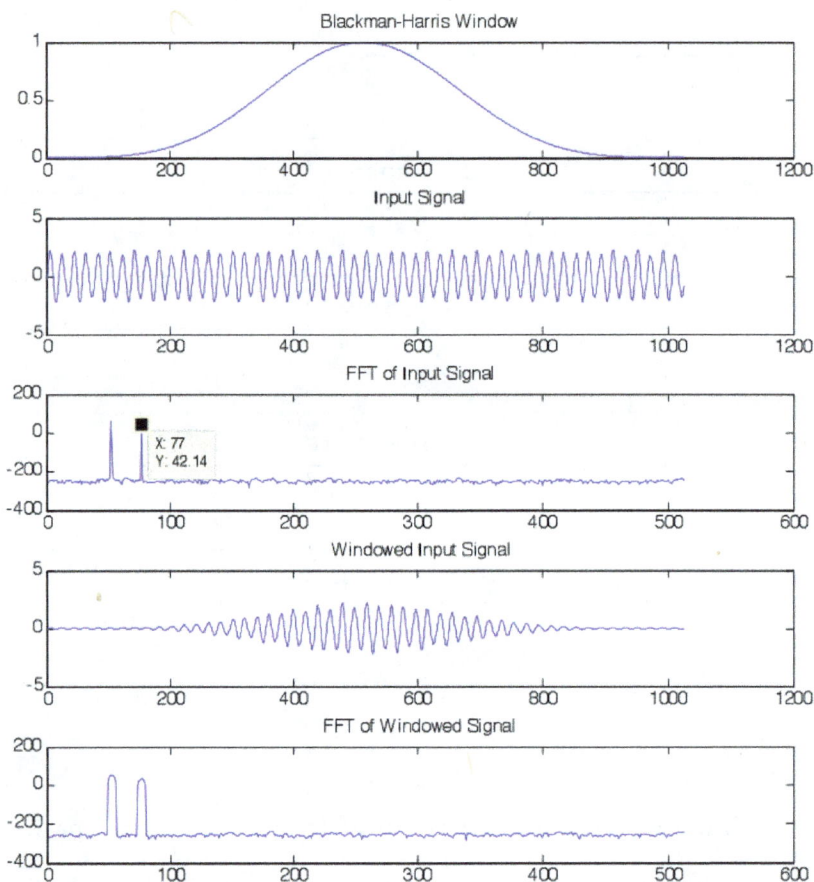

Figure 54. Windowing a signal having no jump discontinuities.

The FFT of the un-windowed signal shows that the signal peak is a startling 270 dB above the noise floor. This suggests about 45 bits of accuracy within the generated signal and the FFT computation.

Because the input signal was "perfect," the application of the window served only to widen the peaks, spreading the energy into adjacent FFT bins. The SNR is still close to 270 dB.

Such a large SNR is not practicable with analog hardware; it can be conveniently generated using double precision floating point computations.

8.12 In-Between Frequencies

When we compute a 128 point complex FFT on 128 input points, the result tells us the amplitude and phase of the 0-cycle (i.e., DC) component, the amplitude and phase of the 1-cycle component, etc. up to the 127 cycle component. The input signal is transformed into the sum of 128 different frequencies that fit within the 128 samples.

What happens if the input signal contains significant energy at, for example, 5.5 cycles? What does the 128-point FFT look like?

First, let's examine the periodic extension: Since the 5.5 cycles ends on a half-cycle, the periodic extension will have an abrupt change in direction. This will affect the FFT results.

The resulting FFT shows most of the energy is bins 5 and 6, while other bins have some energy as well.

Is this correct? The answer is: *yes*. The inverse FFT of the FFT results will match the input signal exactly. The input signal is not 5.5 Hz, it is the periodic extension of a segment having 5.5 cycles.

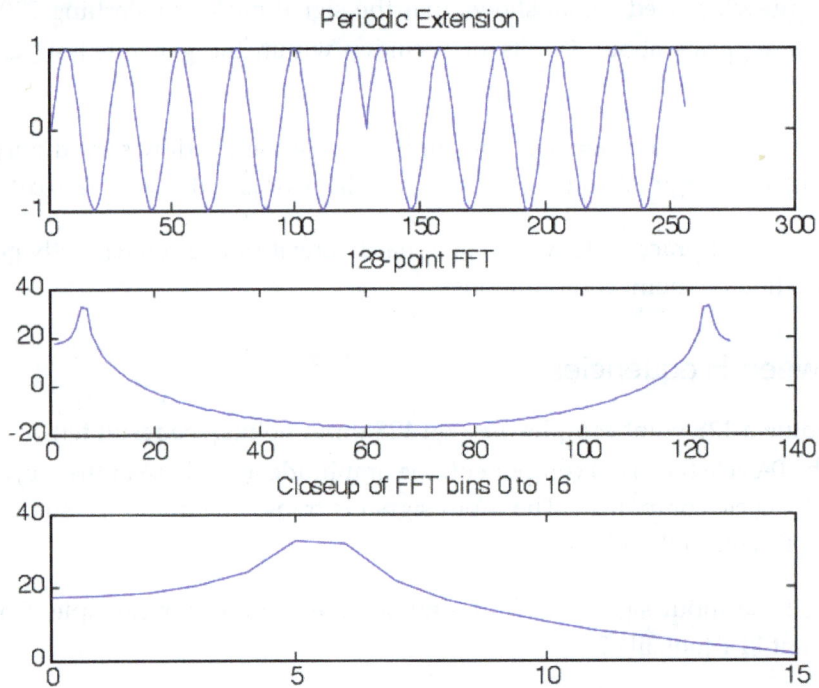

Figure 55. The FFT of an "in-between" frequency.

A related question is, what happens between the 128 points, and does the FFT capture this? The answer to this has two parts: 1) we don't know, and 2) we don't care. When we captured 128 points, we decided that what happens in between the points is of no consequence. Hopefully we collected the 128 points in a way that does not result in aliasing. If we do care about what happens in between the 128 points, we should double the sample rate, collect 256 points, and compute a 256 point FFT.

Though the in-between frequencies have FFT peaks that span multiple bins, the approximate peak frequency can be computed by using quadratic interpolation to find the location of an "in-between" peak.

8.13 The Inverse FFT, DFT

The inverse FFT (IFFT) transforms a vector of frequency domain information (i.e., the spectrum) back to a time domain set of samples.

An inverse DFT is computed by multiplying each complex coefficient in the spectrum by the time domain sequence of the frequency which it represents, and summing these together:

$$S_t = A_0 + A_1 \cdot e^{j\omega 1t} + A_2 \cdot e^{j\omega 2t} + \dots$$

The IFFT is computed in a manner similar to the FFT; the difference being that the complex weights are $e^{j\omega 2t}$ instead of $e^{-j\omega 2t}$.

The IFFT can also be computed using the following method:

- Swap the real and imaginary components of the input vector

- Compute the FFT or DFT

- Swap the real and imaginary components of the result

8.14 FFT, DFT Gain and Bit Growth

For most FFT implementations, computing IFFT(FFT(x)) will generate a vector xN, where N is the length of the vector. The FFT and inverse FFT introduce a gain factor equal to the length of the vector. Software libraries may divide this out, or they may leave it in depending on the library.

A radix 2 FFT for a vector of length N is implemented as $\log_2(N)$ stages of operating on vectors of length N. Each stage involves multiplication by complex values on the unit circle, and adding (or subtracting) pairs of points. The result is that n bit samples into the FFT will grow to $n + \log_2(N)$ bit results. A 1K point radix 2 FFT performed on 12 bit data will produce 22 bit results.

Some FPGA implementations allow scaling at each stage: a 1K point radix 2 FFT performed on 12 bit data will produce 12 bit results.

8.15 Additional FFT, DFT Notes

- Even when the input to the FFT is real, the output will be complex. The exception is: a signal composed of the sum of cosines that fit exactly within the number of samples will produce a real output.

- It is often of interest to display the absolute value (magnitude) of the FFT results. This eliminates the phase. The phase can also be displayed; the phase is important in some applications.

- It is customary to plot the results in dB: $20\log_{10}(\text{abs}(\text{FFT results}))$. Such a display allows you to view and compare peaks having a wide range of amplitudes. In Figure 54 (b), the peak at 76 cycles[25] has a dB value of 42, while the noise floor resides at about -230 dB, for

---

[25] MATLAB numbers indices starting at 1, so point 77 represents FFT output bin 76.

a total SNR of about 270 dB. This is achievable in MATLAB (but not using real ADCs), indicating computation having more than 45 bits of accuracy.

- When the input is real, the real FFT results will be symmetric about the mid point, and the imaginary FFT results will be negative symmetric about the mid point. The second half of the output is redundant, and can often be ignored or eliminated.

- Doubling the input signal amplitude doubles the magnitude of all output bins. Also, the FFT of the sum of two signals is equal to the sum of the FFTs of the two signals. The FFT is a *linear operation*.

- The FFT of random noise is also random, unless the noise has some "coloration".

- The FFT of a single pulse is a flat line; a single pulse in time domain consists of all frequencies.

- The FFT of a flat line is a single pulse at $f = 0$ (that is, a pure DC value).

Exercises

1. We wish to use the Goertzel algorithm to determine the energy content within an audio signal sampled at 8 KHz. The energy content must be evaluated every 50 mSec.

    a. What length should be used for the Goertzel algorithm?

    b. If we wish to compute the complex sinusoid "on the fly", what complex multiplier should we use?

2. A system needs to determine the energy of a signal at 8 specific frequencies over a vector length of 1024 samples. Should the system used Goertzel algorithms, or an FFT?

3. What might you expect as the output of an FFT from a random set of samples?

4. What is the resolution (Hz/bin) of a 1K FFT if the input sample rate is 1 megasamples per second? What is the minimum (other than DC) frequency that can be represented? What is the maximum frequency?

5. Describe the FFT of a single pulse having amplitude 1 at time 0, and 0 at all other times.

References

[1] Kaplan, Wilfred, Applied Mathematics for Engineers, Addison-Wesley Publishing Company, Reading, Massachusetts, 1981, p. 150-151.

[2] Oppenhiem, Alan V. and Shafer, Ronald W., Digital Signal Processing, Prentice-Hall Inc., Englewood Cliffs, New Jersey, 1975, 291-306.

[3] Lyons, Richard G., *Understanding Digital Signal Processing*, Second Edition, Prentice Hall, Upper Saddle River, New Jersey, 2004, pp. 75-82.

Filters

9

In this chapter...

A filter is one of the most common signal processing algorithms. Filters allow some frequencies within a signal to pass, while suppressing other filters. Filters can also be designed to allow all frequencies to pass while altering the phase.

From an algorithm implementation perspective, filters fall into one of two broad categories: finite impulse response filters, and infinite impulse response filters.

Filters are generally difficult to design, requiring software design tools, but relatively easy to implement, requiring dot products. Unlike analog filters, digital filters can be designed to have very precise parameters.

Introduction

A filter is one of the most common signal processing algorithms. Filters allow some frequencies within a signal to pass, while suppressing other filters. Filters can also be designed to allow all frequencies to pass while altering the phase.

From an algorithm implementation perspective, filters fall into one of two broad categories: finite impulse response filters, and infinite impulse response filters.

Filters are generally difficult to design, requiring software design tools, but relatively easy to implement, requiring dot products. Unlike analog filters, digital filters can be designed to have very precise parameters.

Within digital signal processing, a *filter* is a numerical algorithm that, when presented with a stream of input samples representing several frequencies, allows some frequencies to pass while others are reduced or eliminated.[26]

Digital filters perform a similar function to analog filters: circuits consisting of resistors and inductors and/or capacitors that allow some frequencies to pass while reducing the amplitude of other frequencies. However, digital filters can be far more precise than their analog counterparts.

An ideal filter would eliminate all unwanted frequencies, and allow all desired frequencies to pass. However, it is not possible to implement an ideal filter. Actual filters allow desired frequencies to pass. They also allow some undesired frequencies to pass, but at greatly reduced signal levels.

It is not practical (nor is it necessary) to completely eliminate unwanted frequencies using a filter. Instead, the signals need only drive the level of unwanted frequencies down to the level (or slightly below the level) of the noise floor of the system.

9.1 A Simple Filter

The *moving average* of a data set replaces each point in a vector with the average value of the last k input points. This is often used to "smooth" a data set. For example, if an input vector has these values, in order of increasing index:

> 1 3   8   4   12   6   8   7   11 1

then the moving average of four samples generates a new vector where each point is the average of four samples:

---

[26] Where "eliminate" often means to make so small that they do not matter.

(1+3+8+4)/4;   (3+8+4+12)/4;   (8+4+12+6)/4;   (4+12+6+8)/4   ... =

16/4; 27/4; 30/4; 30/4; 33/4; 32/4; 27/4

Figure 56 shows a random signal, and the result of applying the moving average filter to the signal. The second plot in the figure shows significant smoothing. The third plot shows the result of applying the filter a second time to the data; the contours become even smoother.

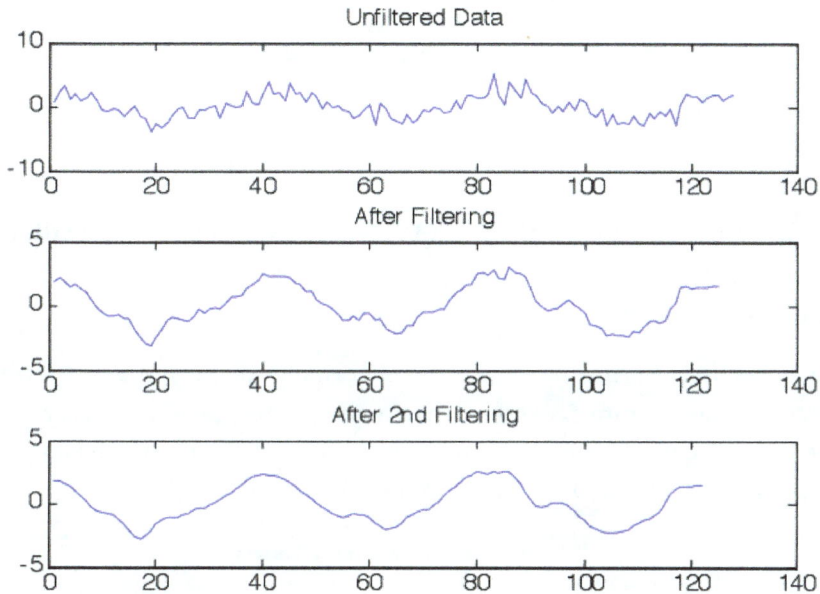

Figure 56. Application of a moving average filter.

This is an example of a low-pass filter: it allows low frequencies to pass through, while reducing the amplitude of higher frequencies. The data flow diagram for this filter is shown in Figure 57. The z^{-1} blocks represent delays of one sample. This filter requires a delay buffer, or state buffer, of 3 samples, to hold the prior 3 input samples.

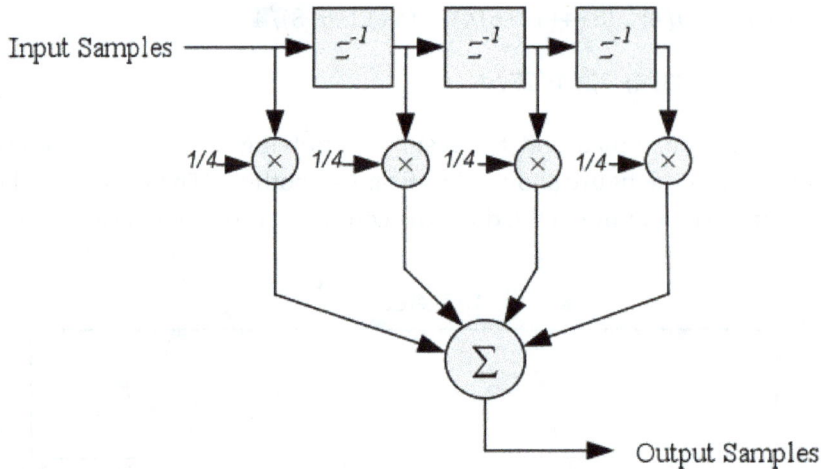

Figure 57. Signal Flow Diagram for 4-point Moving Average Filter.

9.2 Frequency Response

All filters have a specific *frequency response*. This is normally displayed as a graph of dB (y-axis) versus frequency (x-axis). Figure 58 shows the frequency response of a very precise low-pass filter. This filter allows frequencies of 40 Hz and below to pass, virtually unchanged in amplitude. Frequency components greater than 50 Hz are reduced in amplitude by 70 dB.

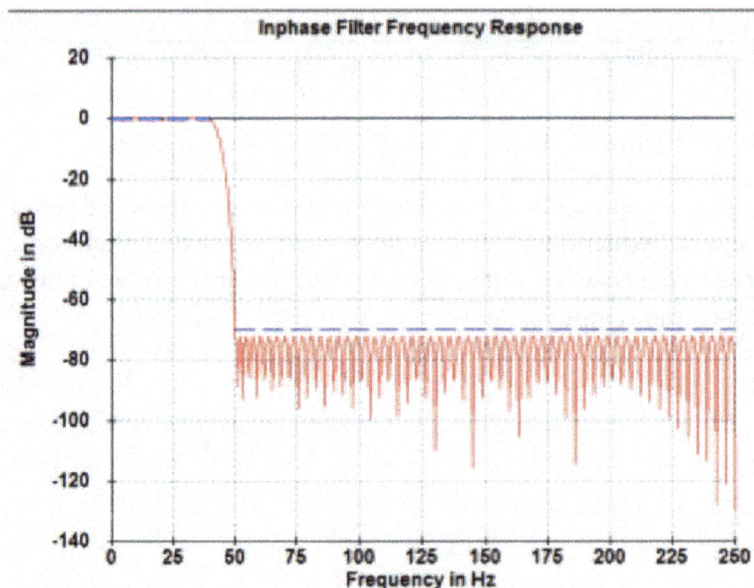

Figure 58. Frequency response of a precise low-pass filter.

Figure 59 shows the frequency response for the 4-term moving average filter. At 256 samples per second, this filter will eliminate[27] components at 64 and 128 Hz, and allow frequencies of 0 to 5 Hz to pass through almost unchanged. All other frequencies will be reduced according to the shape of the frequency response curve.

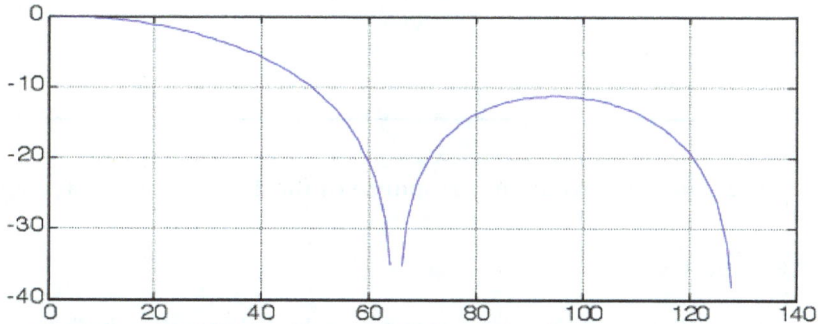

Figure 59. Frequency response for a 4-term moving average filter.

At 256 samples per second, a frequency of 64 Hz has a sequence such as 5, 0, -5, 0, 5, 0. The average value of any four terms of such a sequence will be zero. A frequency of 128 Hz has a pattern such as -3, 3, -3, 3. The average value of any four terms of a sequence having this form will also be zero.

9.3 Filters and Complex Samples

Filters can operate on real or complex sample streams. When operating on a real stream, the filter frequency response plots start at 0 and extend to Fs/2. The frequency response is sometimes called the *one-sided* response.

When operating on a complex stream, the filter frequency response starts at –Fs/2 and extends to Fs/2. The response is symmetric about zero Hz. The response plot is sometimes called the *two-sided* response. Figure 60 shows the two-sided response for the moving average filter. With this plot, we see that the filter will eliminate complex frequency components of 64 and -64 Hz, 128 and -128 Hz.

---

[27] While filters greatly reduce the amplitude of many frequencies, this filter will totally eliminate frequencies of 64 and 128 Hz.

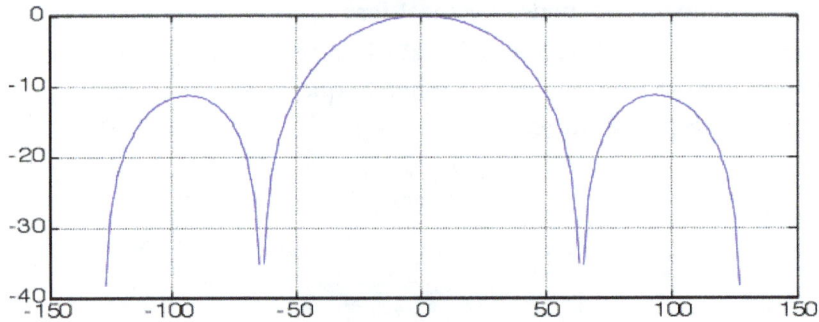

Figure 60. The two-sided frequency response of the 4-term moving average filter.

9.4 Some Filter Terms

A *low-pass* filter allows low frequencies to pass while eliminating high frequencies.

A *high-pass* filter allows high frequencies to pass while eliminating low frequencies.

A *bandpass* filter allows frequencies in the middle of the spectrum to pass, while eliminating high and low frequencies.

A *notch* filter allows all frequencies to pass except for a very narrow range of frequencies.

The *passband* of a filter describes what frequencies are allowed to pass. It is normally measured from the maximum value within the passband to the points where the response drops by 3 dB. If the passband of a filter is 70 to 80 MHz, the filter bandwidth is 10 MHz.

The *bandwidth* of a filter is the highest frequency in the passband minus the lowest frequency.

The *stopband* of a filter describes what frequencies are greatly reduced.

The *stopband attenuation* of a filter describes the level to which the filter drives the unwanted signals. This is usually expressed in decibels.

The *transition width* describes the difference in frequency between where the passband ends and the stop band begins. This is also called the "roll-off".

The filters whose frequency responses are shown in Figure 58 and Figure 59 are low-pass filters that have the following characteristics:

Parameter	Precise Lowpass	Moving Average
Passband	0 to 40 Hz	0 to 30 Hz
Passband (2-sided)	-40 to 40 Hz	-30 to 30 Hz
Bandwidth	40 Hz	30 Hz
Bandwidth (2-sided)	80 Hz	60 Hz
Stopband	$f > 50$ Hz	$f > 64$ Hz
Stopband Attenuation	> 70 dB	11 dB
Transition Bandwidth	10 Hz	30 Hz

9.5 Filter Algorithms

Filter algorithms fall into two classes: Finite Impulse Response (FIR) and Infinite Impulse Response (IIR) filters.

For now, ignore what these class names might mean. A simple description is this: a FIR filter is implemented by computing the dot product between the last K inputs, and K fixed weighting values, called filter taps. That's it for a FIR. The magic of what type of filter it is (low pass, high pass, etc), where the cutoff frequency is, what the stop band attenuation is, etc. is in the number and values for the filter taps.

A simple description of an IIR filter is this: it is the dot product of the last K input samples with a set of taps, plus the weighted sum of the last N output samples of the filter. The output of an IIR filter depends on both the last K inputs, and the last N outputs that it generated.

9.6 FIR Filter

A FIR (finite impulse response) filter is a specific type of digital filter. A FIR filter is called "finite impulse response) because when the input is an impulse (an impulse is a nonzero sample followed by zeroes), the output will be nonzero for a finite number of samples, followed by samples having zeroes. The filter response "dies out" to zero after a finite number of samples.

A FIR filter generates each output from the weighted sum of the last N inputs, where the number of weights (N) and their values determine if the filter is low pass, high pass, or any other form. The weights are also called *coefficients* or *taps*. Normally a filter design tool is used to generate the weights from an input specification.

A plot of the filter taps for the 40 Hz low pass filter illustrated above is provided below. Note that the tap values do not look anything like the frequency response plot.

9.7 FIR Filter Taps

Filter taps are computed from a filter specification. Because of the number of parameters and complexity in computing taps, software tools are employed to generate taps. The specification describes the following parameters:

Parameter	Description
Type	High pass, low pass, etc.
Pass Band Frequency	
Stop Band Frequency	
Transition Region Width	pass band – stop band
Stop Band attenuation	dB
Passband Ripple	How flat is the passband, in dB
Desired number of taps	Often specified as NTE

Not all parameters are important for all applications. For example, we might want need a small number of taps. A lower number of taps requires less computation for each output sample. To achieve this, a designer may widen the transition region width of the filter.

The *passband ripple* describes how flat the frequency response is within the passband. Some frequencies within the passband will pass through with slightly more energy than others; the high points within the passband indicate this. Likewise, the low points within the passband correspond to frequencies that will be "slightly attenuated".

By the way, an engineer might be tempted to put two of the same digital filters in a row to construct a "better" filter. This will double the stop band attenuation, and improve the rolloff (the slope of the transition region). Unfortunately, it also doubles the passband ripple.

For modem signals, values of 0.1 dB or lower for passband ripple may be needed. For audio signals, values of up two dB or so are not noticeable to the human ear unless you can also listen to the "ideal" audio at the same time, for comparison.

The 40 Hz low-pass filter above was designed using an "equiripple" design tool – ripples within the passband are equal, and ripples within the stop band are equal. The filter requires 134 taps to achieve the desired passband, stopband attenuation, and transition region width. A plot of the filter tap values is shown below.

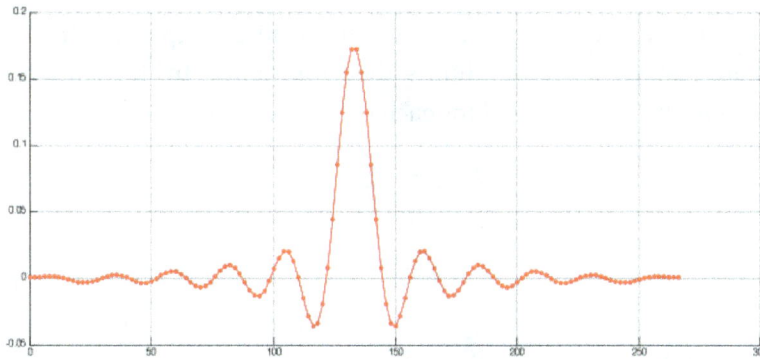

Figure 61. The impulse response for the low pass filter.

FIR Filter Impulse Response

When a FIR filter is presented with an impulse input signal (that is, a 1 followed by all zeroes), the resulting output signal is the tap values, in reverse order.

Number of Taps

The number of taps is often traded off against the width of the transition region. These two values are inversely related: a short transition region results in a large set of filter taps; a wide transition region results in a smaller number of taps.

The delay through a filter is also a function of the number of filter taps. The output of a filter corresponding to a specific input occurs after (ntaps/2) have been processed.

Tap Symmetry

A FIR filter having symmetric taps has a *linear phase* response. That is, all frequencies within the passband are delayed by the same amount of time. This is important for signals that use phase information to convey information. It is less important for audio signals, since phase relationships cannot be discerned by a listener.

Figure 62 shows the phase response for the lowpass filter. This shows that within the passband, the phase is a linear function of the frequency. This filter has a linear phase response.

Phase linearity implies that the output of a filter will look like a delayed version of the input signal, provided the input for those frequencies that are within the filter's passband. A square wave is the sum of multiple frequencies; if most of these frequencies are within the passband of a linear phase filter, the output will be a square wave. If the filter does not have a linear phase, the output will no longer be a square wave, since the various harmonics will be shifted around.

FPGA filter implementations exploit tap symmetry by adding samples having the same tap value to form intermediate results; the tap value is then multiplied by this intermediate sum. This halves the number of multiplies required for each output sample.

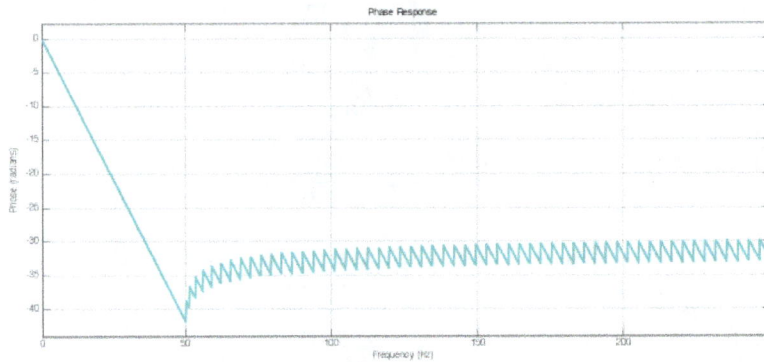

Figure 62. Phase Response of the Low-Pass Filter.

Real versus Complex Taps

Filters need not always have real taps, but often they do.

When applied to a complex signal, a filter having real taps can be applied to the real and imaginary inputs separately, to generate the real and imaginary components of the output.

When a filter having complex taps is applied to a complex signal, the multiplies and adds within the filter must be complex operations; it is not possible to separately process the real and imaginary components in this case.

Tap Values

The tap values can be multiplied or divided by a constant without changing the shape of the filter, provided the new filter taps still fit within the "bit width" used within the implementation. This has the effect of adding gain to the filter (or attenuation). It pushes all points of the frequency response curve up or down without changing the shape of the curve.

Filter Gain

The taps of a filter can be multiplied by a real constant. This does not affect the shape of the frequency or phase response curves, but it does affect the dB scale. A filter can perform filtering while increasing or decreasing the signal level.

Often, filter generation tools generate floating point taps having values between -1.0 and 1.0. The taps are often scaled so that the filter provides 0 dB of gain within the passband. Such values are normally scaled to integers before implementing the filter in an FPGA.

One can, for example, multiply the filter taps by 32767, and round to nearest, to obtain 16 bit coefficients.[28] The filter output may need to be shifted right by 15 bits and rounded to obtain a 0-gain result.

Often, however, the largest filter tap is still a small fraction. For the lowpass filter, the largest tap is 0.1720. Other taps are very close to 0: taps on either end are 0.00029. When scaled by 32767, these values do not require a full 16-bit range. In order to provide greater accuracy, the filter coefficients can be scaled by 32767/0.1720. Doing so reduces, but does not eliminate, imperfections introduced when the filter coefficients are converted to integer values. The filter output can then be scaled by the inverse value to provide 0 dB gain within the passband.

9.8 Cascading FIR Filters

FIR filters may be cascaded (tied end to end) to achieve an improved overall response. The moving averaging filter was applied twice to the data in Figure 56; this improved the smoothing effect.

When applying the same filter twice, each filter must have its own separate history buffer. Also, while the stopband attenuation and transition regions combine in a manner that is helpful, any ripple within the passband doubles. For this reason, when multiple filters are combined, the filters are often designed with different parameters, so that overall passband ripple meets system design constraints.

9.9 The DFT as a FIR Filter Bank

Since a FIR filter is the dot product between a set of taps and an input signal, and since a single bin of the DFT (or FFT) is the dot product between an input signal and a complex sinusoid, we can consider the DFT of a vector of length N to be a set of FIR filters. Figure 63 shows the linear frequency response for the "filters" associated with bins 3, 4, and 5 for a 128-point FFT. All filters

---

[28] Coefficients are 2s complement values

have the same response; they are simply shifted to other frequencies. Figure 64 shows the same response curve using a dB scale.

Note that a single filter "peaks" at its center bin value. Note also that all of the filters have a response of 0 at all other integer frequency values.

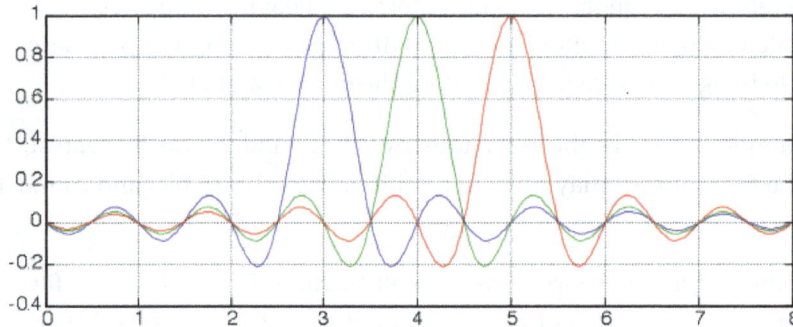

Figure 63. Frequency Response for DFT Bins 3, 4, 5.

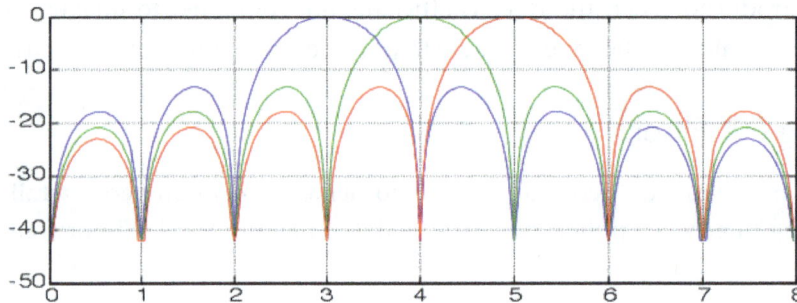

Figure 64. Frequency response (dB) for bins 3, 4, 5.

Of course, a filter is normally applied to all points in a signal, successively shifting single points into the filter and shifting older points out. An FFT or DFT would be applied to a set of N points, and then several points would be shifted into the next input data set for the FFT.

9.10 IIR Filters

An IIR (infinite impulse response) filter is a specific type of digital filter. An IIR filter is called "infinite impulse response" because when the input is an impulse (an impulse is a nonzero sample followed by zeroes), the output may never die out.

In fact, the impulse response of an IIR filter could remain the oscillate forever, or diminish exponentially, or even grow exponentially. It may also die out for *some* inputs signals.

The determination of whether a filter is FIR or IIR is *not* done on the basis of seeing how it responds; rather it is determined by the architecture of the filter. An IIR filter generates an output point based on the weighted sum of the last K input points, *plus the weighted sum of the last M output points that it generated.* It is the presence of feedback terms in the filter that causes it to be classified as an IIR filter. It is also this feedback, in combination with the filter taps, that create situations where an input that is nonzero for a small number of samples may result in nonzero outputs forever.

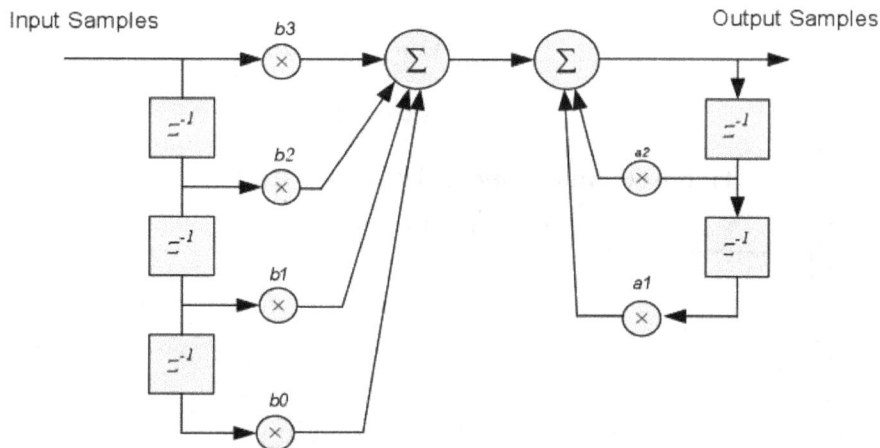

Figure 65. A Typical IIR Filter Structure.

IIR filters can be designed to form lowpass, high pass, bandpass, or notch filters. The design is trickier, and the filters must be analyzed for stability. IIR filters do not exhibit linear phase delay over the entire spectrum, but they may do so over narrow ranges.

IIR filters are more sensitive to rounding. Coefficients are normally computed using tools that employ high precision. Rounding coefficients to fixed-point values slightly alters the filter behavior. A filter that has good behavior as indicated by an analysis and design tool may fail when coefficients are rounded to fixed point values having less accuracy.

IIR filters often exhibit phase characteristics that make them unsuitable for many applications.

IIR filters are generally more efficient than FIR filters, in that filters can be realized using fewer multiplies and adds. But the difficulty in generating a good, stable filter that is well behaved means that FIR filters are often preferred.

Exponential Averager

The exponential averager is a simple IIR filter. Like the moving average filter, the exponential averager "smooths" the data. The impulse response is an exponential decay. The rate of decay is

143

determined by α, which has a value between 0 and 1. Values closer to 0 result in a slower decay. The signal flow diagram for the exponential averager is shown in Figure 66.

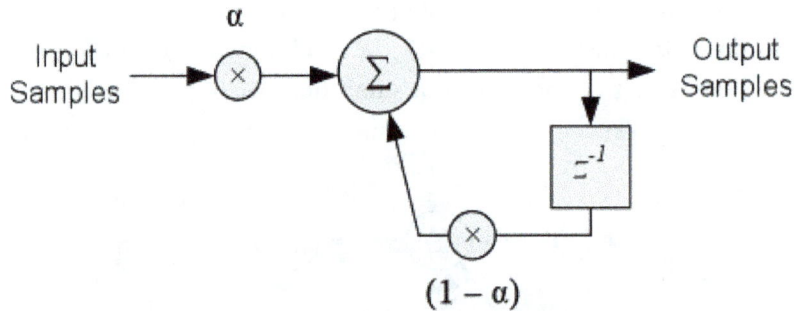

Figure 66. An Exponential Averager IIR Filter.

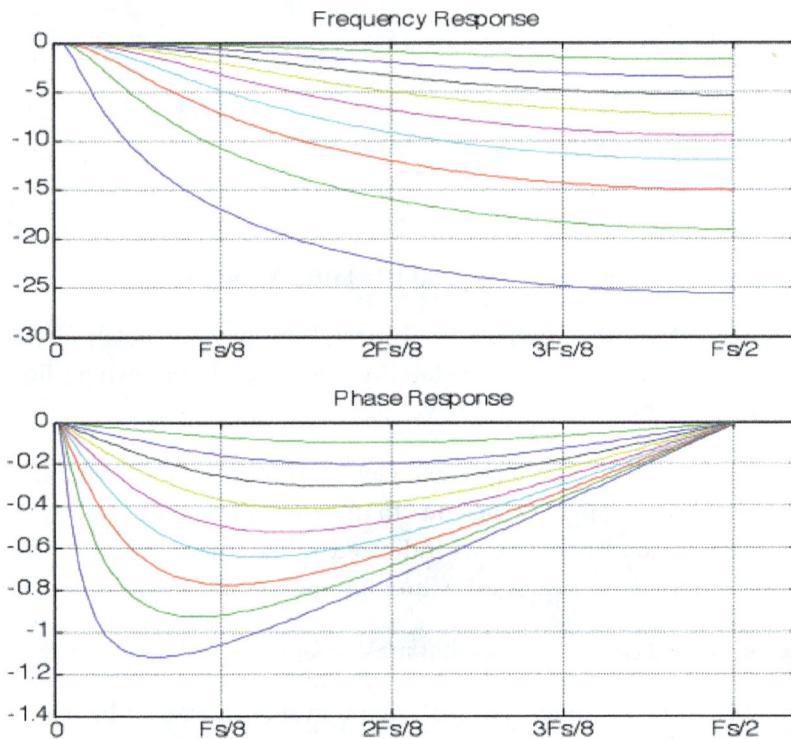

Figure 67. Frequency and Phase Response for α values of 0.1 to 0.9.

9.11 Six Views of the FFT

Having discussed filters, dot products, convolution, and correlation, we can now view the FFT in multiple ways. Remember that the FFT is a more efficient implementation of the DFT, so these views apply to both the DFT and FFT. The DFT can be viewed in any of the following equivalent ways:

1. The DFT of a set of samples is a set of complex coefficients $A_0...A_{n-1}$, that, when multiplied by complex sinusoids of frequencies 0, 1, ... n-1 and added together, recreate the original signal. A_0 is the DC component, A_1 is coefficient for the fundamental frequency, and all other coefficients represent the magnitudes and phases of harmonics. Therefore, the sum of the sinusoids also generates the periodic extension of the original sample set.

$$A_0 + A_1 e^{-jwt} + A_2 e^{-j2wt} + \ldots A_{n-1} e^{-j(n-1)wt}$$

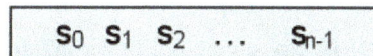

2. The DFT of an n length vector is the product between an input vector (the signal), and a square $n \times n$ matrix, where columns of the vector are complex exponentials of frequencies 0..n-1.

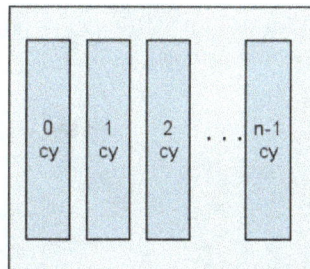

The inverse DFT is obtained by multiplying the A_n coefficient vector by another matrix, having columns with negative frequencies.

3. The DFT is the result of applying the Goertzel algorithm N times, of frequencies (0..n-1), computed on a signal vector.

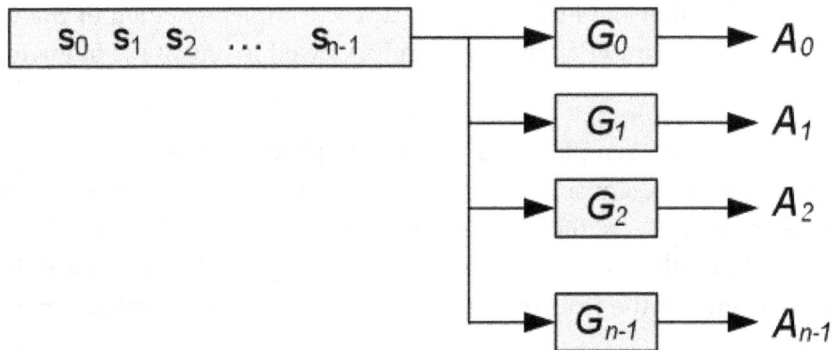

4. The DFT is a set of correlations between n different complex exponentials of frequencies 0..n-1 and an input signal vector.

5. The DFT is a set of n FIR filters (convolutions), where the frequency response of each filter is a single peak, and the impulse response is a complex exponential, for complex exponentials of 0..n-1.

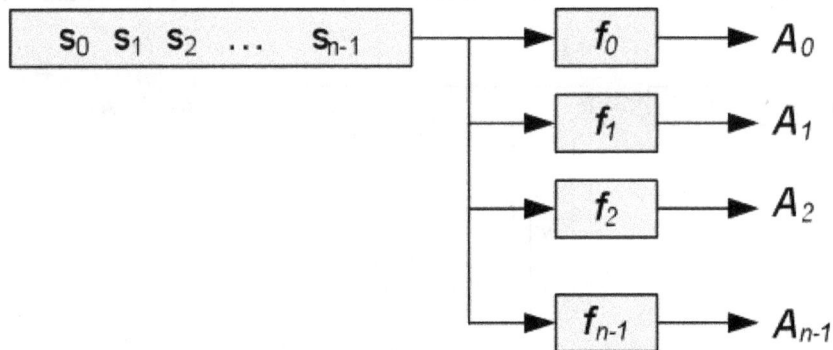

6. An output bin n of the DFT is obtained by downconverting the input signal using a complex exponential of frequency e^{-jwnt}, and summing. This gives n × the DC value of the downconverted signal.

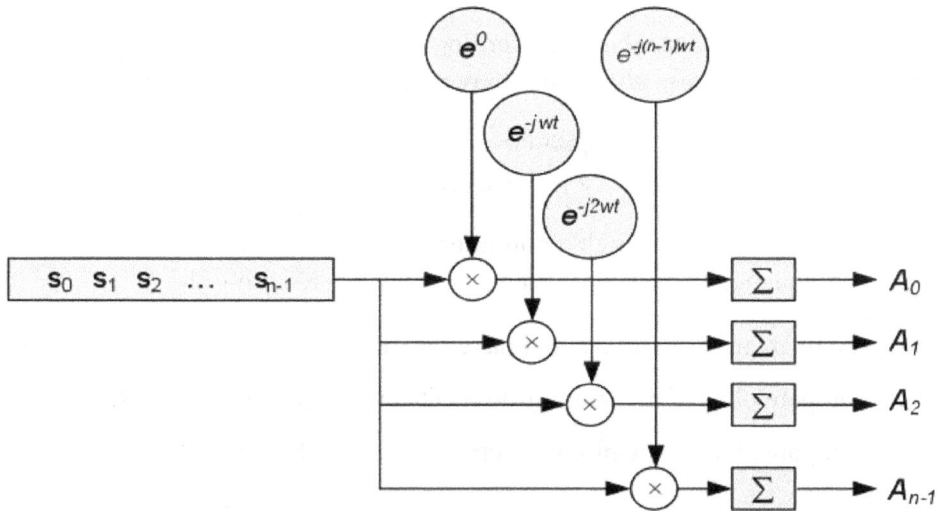

All of these views consist of the dot product between the input set of samples, and complex sinusoids of frequencies 0 to n-1.

Exercises

1. A filter having linear phase introduces a constant delay for all those frequencies which it passes. Why is this phenomenon called linear phase?

2. If human hearing is not capable of discerning phase, why would one employ linear phase filtering, when a nonlinear phase filter may require fewer operations per sample?

3. A square wave is passed through a linear phase, low-pass filter having a passband of 9 times the frequency of the square wave. Describe the resulting signal.

4. A square wave is passed through a non-linear phase, low-pass filter having a passband of 9 times the frequency of the square wave. Describe the resulting signal.

5. Assume the filters in exercises 3 and 4 have the same passband frequency response.

    a. Compare plots of the magnitude spectrum for each filter output.

    b. Compare plots of the phase spectrum for each filter output.

Decimating and Interpolating

10

In this chapter...

The sample rate at which a signal is measured may not be to our liking. The sample rate is determined by hardware, and may be higher than needed. Or, it may be lower than needed. Or, if we are going to demodulate a 9600 baud modem signal, we might want the sample rate to be a multiple of 9600.

Changing the sample rate is accomplished through decimating, interpolating, or combining the two together. The goal is to change the sample rate but represent the exact same signal within the spectrum (that is, to avoid introducing noise during the process.)

Filters, when combined with decimators, interpolators, or both, are called rate change filters, or multi-rate filters.

10.1 Decimation

Decimating a sample stream lowers the sample rate by discarding samples and keeping only one sample for every N input samples. Decimating by two, for example, discards every other point. This lowers the sample rate: the output sample rate of a decimate-by-N process will be $1/N$ times the input sample rate.

Mathematically, if a signal can be processed at a lower sample rate and still yield the correct results, then decimation can be employed in order to reduce the sample rate. Processing signals at a lower sample rate requires fewer operations per second, freeing up FPGA resources and/or CPU cycles to perform other operations.

10.2 Requirements for Decimation

The goal of decimation is to reduce the sample rate but keep the same spectral content. Before a sample stream is decimated, one must ensure that the signal consists of frequency components that can be represented at the decimated sample rate without aliasing. The signal should contain no frequency components greater than or equal to one-half of the decimated sample rate.[29] If this condition is not met, frequencies beyond this limit will alias to new locations within the output spectrum. Once aliased, the components cannot be separated from components that were not aliased.

The figure below shows an input signal that *can* be decimated by two. The spectrum of the original signal has no frequency components above the "halfway point" of 256 in this case. Decimating spreads the signal out (note the change in x-axis), in both the time and frequency domains.

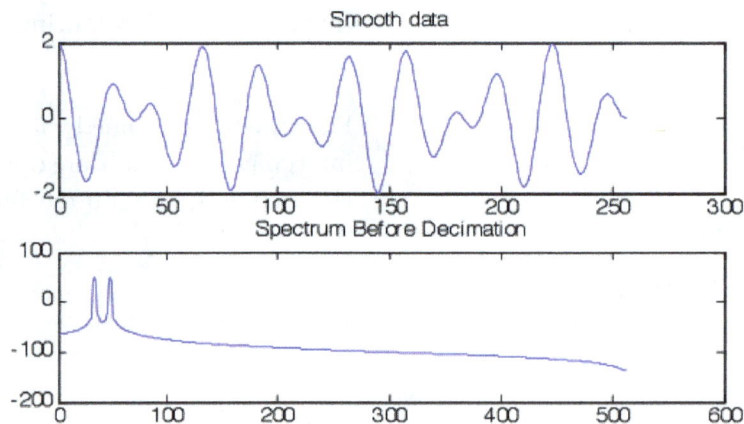

---

[29] Actually, the signal should have a bandwidth less than Fs/2 for real signals, or Fs for complex signals, in order to prevent aliasing from destroying information.

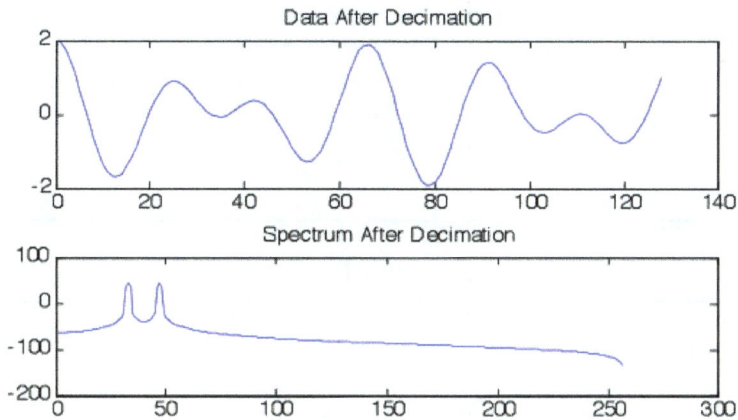

Figure 68. Decimating a band-limited signal by 2.

The signal below contains a high-frequency component. After decimating by two, the resulting sample rate will not be greater than twice the highest frequency component. When decimation "spreads out" the x=axis, the high frequency "wraps back" into the spectrum. The input signal showed peaks at 32, 46, and 310; the decimated signal shows peaks at 32, 46, and 202. The 312 "wrapped" to (512 – 310) = 202.

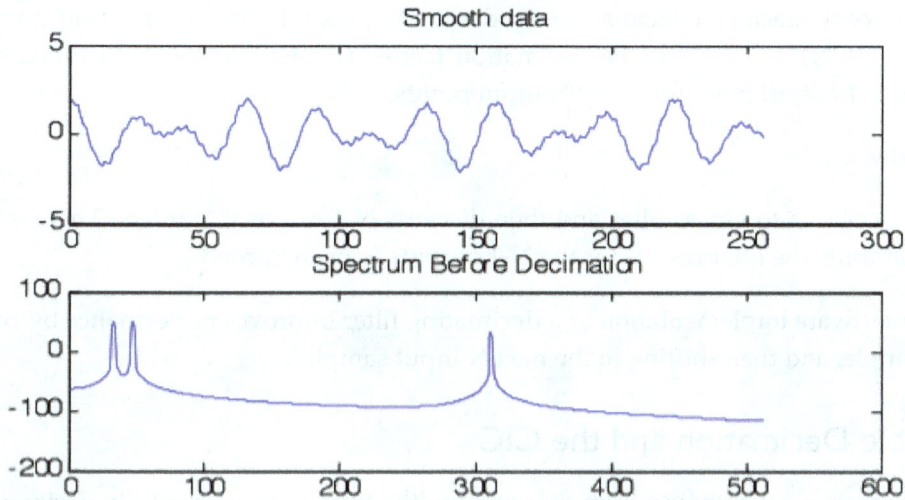

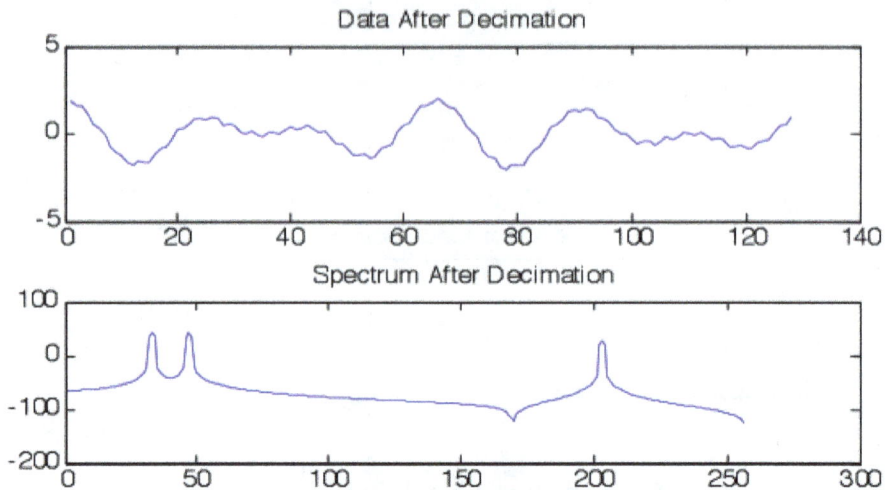

Figure 69. Decimating a non-band-limited signal by 2. Note the aliasing.

In this "clean signal" example, we had prior knowledge that the signal contained a 310 Hz component, and we know exactly where it moved as a result of aliasing. But filters and decimation are often used on signals where we do not have prior knowledge of the content.

In order to prevent aliasing, a decimator includes a low-pass filter having a cut-off frequency no greater than $F_s/(2N)$, where N is the decimation factor. This is followed by a functional block which provides 1 output point for every N input points.

Making It Faster...

A decimator applies a low-pass filter and then discards N-1 of every N filters. There is therefore no need to compute the filter results for the N-1 outputs to be discarded.

An FPGA or software implementation of a decimating filter improves performance by computing an output sample, and then shifting in the next N input samples.

10.3 Variable Decimation and the CIC

In order to achieve good response from a decimator, the filter must be carefully designed to have a flat passband, good stop band suppression, and a transition region that is narrow compared to Fs/(2d), where d is the decimation factor. For large values of d, such a filter may be difficult to construct. Furthermore, if the filter is realizable, it is likely to have a large number of taps requiring a significant dynamic range.

For large values of d (say, d > 4), it may be necessary to decimate using multiple stages: decimate by 2, then decimate by 2 again, etc. This approach is somewhat expensive, since each stage requires multiple taps, registers, delay lines, etc. Also, the total decimation factor achieved by this approach is limited to composite numbers having a large number of small factors. Decimating by 16 is possible with multiple stages, decimating by 17 is not, since 17 is prime.

For high values of d, a better approach exists. The approach uses a cascade integrator-comb (CIC) filter, followed by one or two specially designed FIR filters.[30]

A CIC filter consists of a set of integrators, followed by a decimation stage, followed by a set of comb filters. Details of a CIC filter can be found in Lyons and Harris. Rather than describe CIC filters here, we will simply summarize.

The CIC filter offers the following advantages:

- Integrators and combs require no multiplications; they require only adders and subtractors. CIC filters are therefore easy to implement and they operate at high speeds using a minimum amount of logic.

- The frequency response has approximately the same shape for any value of d. CIC filters can therefore decimate by an arbitrary value.

- The decimation value, d, can be changed dynamically. Filter "taps" need not be recomputed.

However, a decimating CIC filter has the following disadvantages:

- Only a small portion (1/4th or less) of the resulting passband is useable, due to aliasing of other components that increases for frequencies farther away from 0.

- The usable portion of the passband is not flat; it has a slight curvature (droop).

- The filter has significant passband gain, and requires very wide adders and registers.

These disadvantages are easily addressed. The first two disadvantages are addressed by following the CIC filter with FIR filters that band-limit and decimate the CIC output, eliminating the unusable portion of the CIC result. The FIR filter is designed with an inverse curvature to compensate for the passband curvature (in order to compensate for the CIC passband droop). Such a FIR filter is called a compensation filter. Finally, the wide adders and registers are conveniently implemented within FPGAs.

---

[30] CIC filters can also be used for interpolation.

CIC filters are often implemented in multiple stages. Each stage provides about 13dB of lobe suppression for lobes located at about 1.5Fs. Increasing the number of stages reduces the amount of out-of-band energy that aliases back into the passband. An increased number of stages also increases the filter gain, and the passband droop. The zeros, however, remain at the same location.

The figure below shows the "unwrapped" frequency response for decimating CIC filters having 3, 4 and 5 stages. The x-axis is frequency, but frequencies above 0.5 times Fs will alias back into the passband. Note that lobes are 40 dB down.

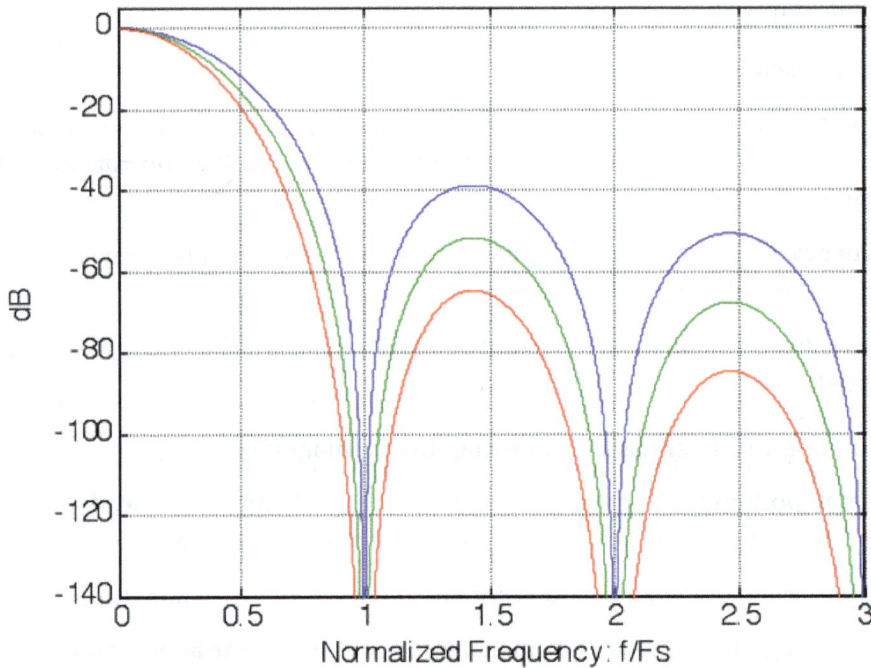

Figure 70. 3, 4 and 5-stage CIC filter responses (unwrapped).

When the spectrum is plotted with the wrapping, we see that high frequency inputs will alias back into the response. The figure below shows the wrapping for a 3-stage CIC filter. This particular 3-stage CIC is not very discriminating, since aliasing components are as high as -50dB at 0.12*Fs, and worse for higher frequencies.

154

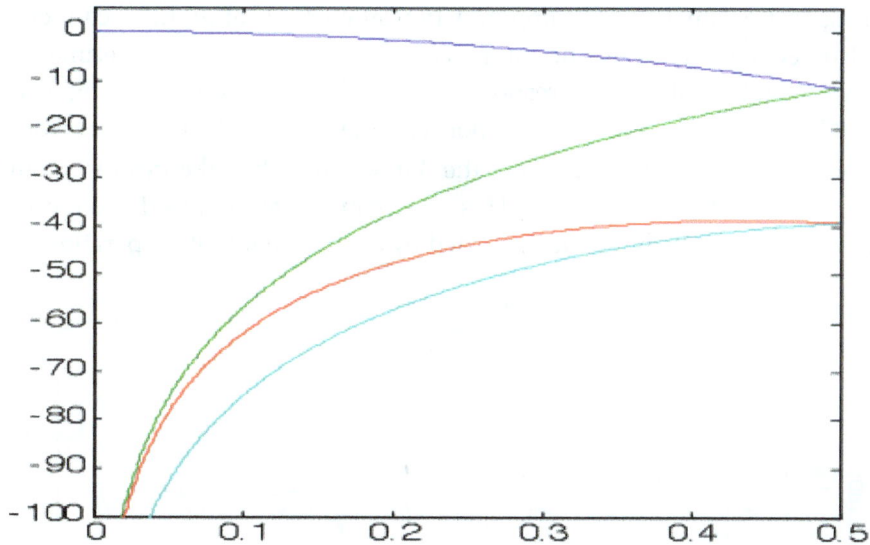

Figure 71. Aliasing components for 3-stage CIC filter.

The figure below shows aliasing components for a 5-stage decimating CIC filter. If we employ additional FIR filters to restrict the output to, say, less than 0.125 of the CIC output, aliasing components are approximately 80 dB down from the central lobe.

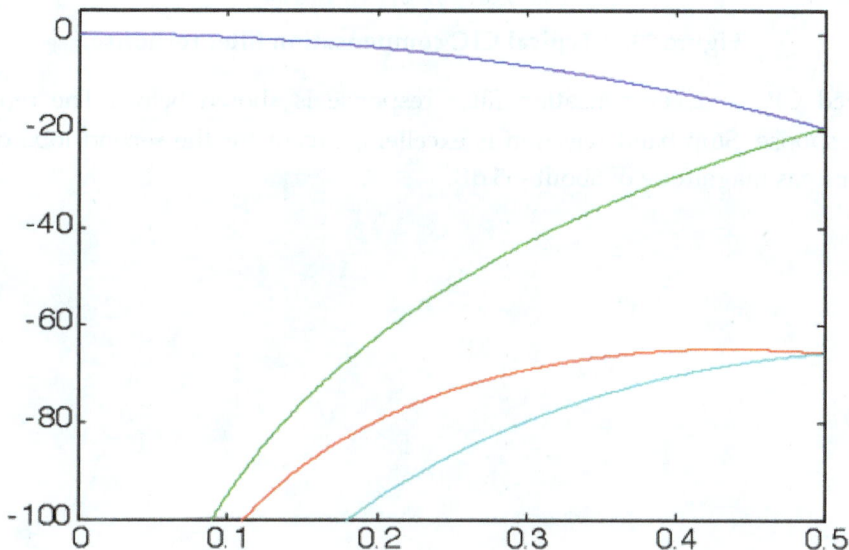

Figure 72. Aliasing components for 5-stage CIC filter.

Since it is desirable to use only that portion of the CIC output near zero in the frequency domain, further filtering and decimation is required. A typical compensation filter corrects the passband droop, and also decimates by two. The compensation filter is a low-pass, decimate by 2 filter that has the inverse curvature of the CIC response within the passband. The compensation filter for the 5-stage CIC filter has the following response. Note the slight upward curvature within the passband. This filter was designed with the knowledge that the results would be further decimated by two by another FIR filter. This filter can therefore afford a wide transition band between 0.25 and 0.75Fs, which will be discarded by a subsequent filter operation.

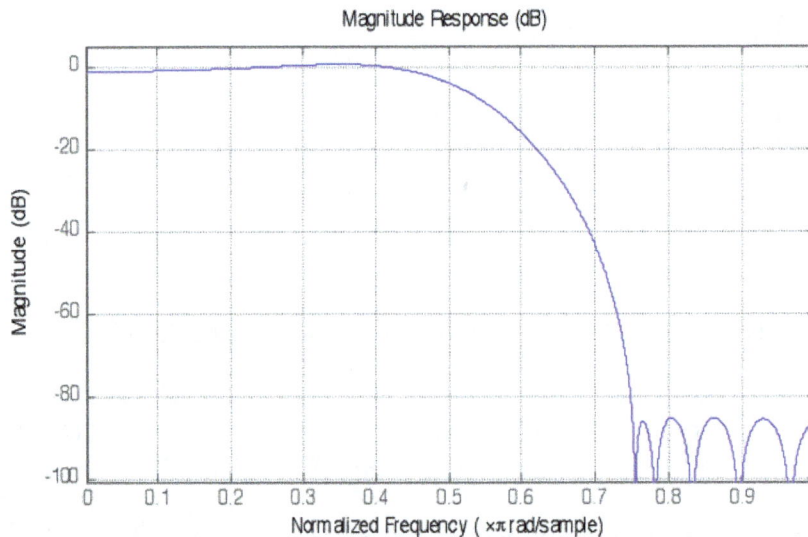

Figure 73. A typical CIC compensation filter response.

The combined CIC and compensation filter response is shown below. The red curve is the composite response. Stop band rejection is excellent, except for the second lobe out of the CIC and CFIR; this has magnitude of about -55 dB.

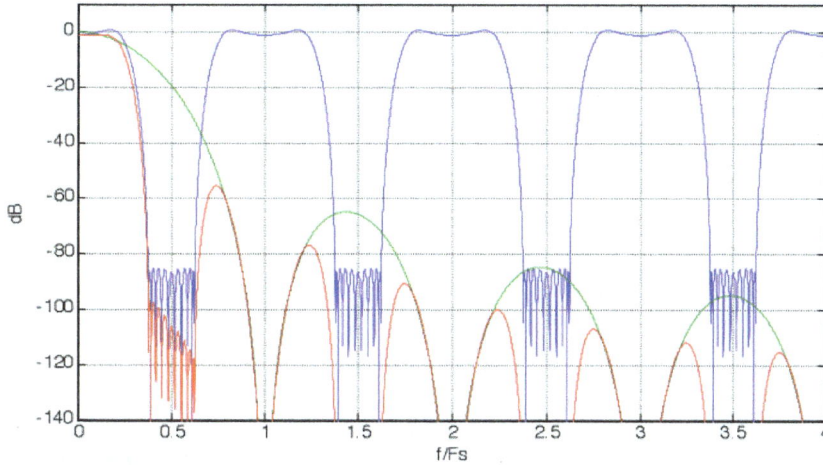

Figure 74. Combined CIC and compensation filter response (red).

The figure below shows a close-up of the frequency response from 0 to Fs. The composite response is flat, and stopband suppression is good except for the lobe at 0.75 Fs.

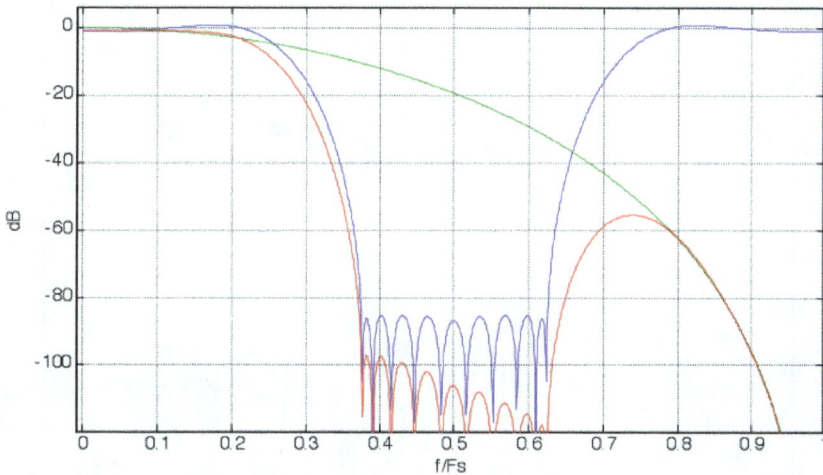

Figure 75. Combined CIC and compensation filter response detail.

A final FIR filter having flat passband response and decimation by two cleans up the final response. The filter below was designed so that the transition region begins at 0.4Fs.

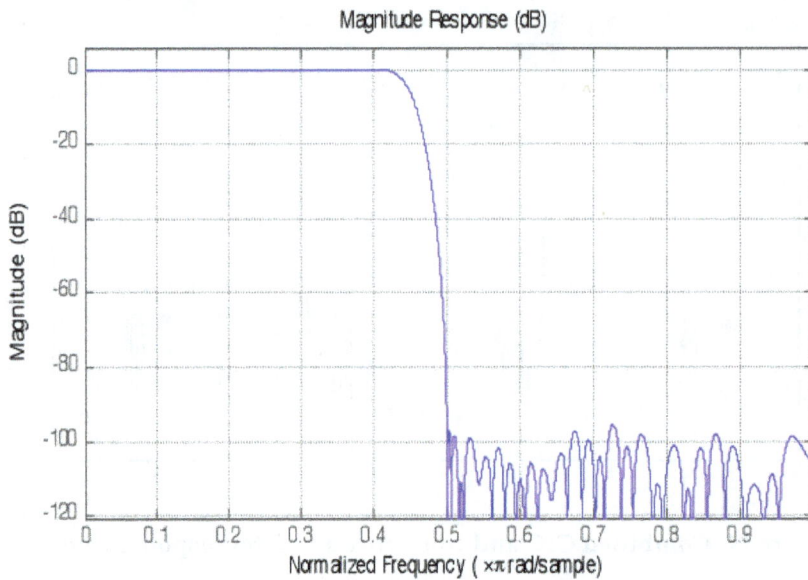

Figure 76. A final decimate by 2 FIR.

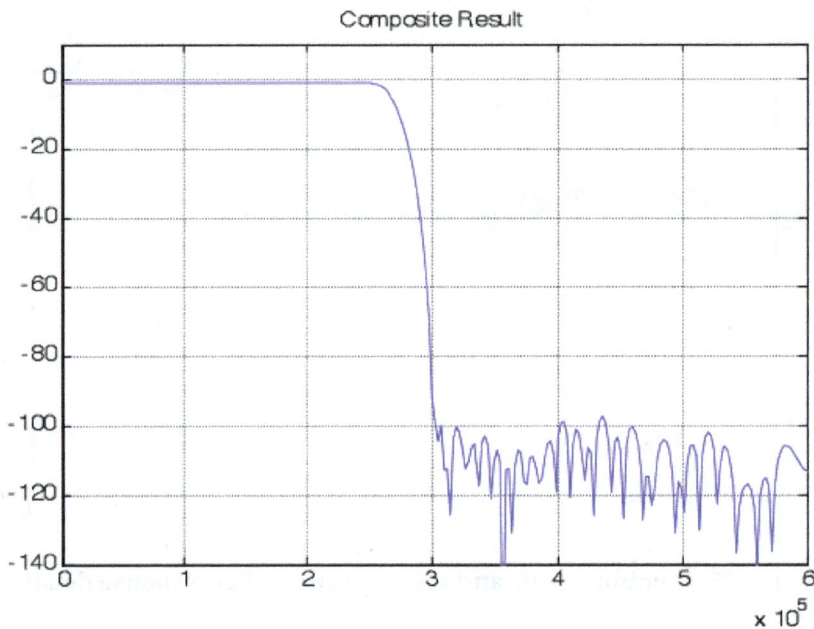

Figure 77. Composite CIC, compensation, and final FIR filter response.

10.4 Interpolation

Interpolation increases the sample rate of a signal. As the name implies, interpolation "fills in" new samples between existing samples within the sample stream, increasing the sample rate. For example, a 2x interpolator would provide an output sample rate of twice the input sample rate, *representing the same input signal* (that is, having the same spectral components) with twice as many samples per unit time.

At first thought, it might seem that a 2x interpolator should insert new samples between the existing samples by computing the average value between each pair of original samples. When analyzed within the frequency domain, this approach performs poorly, introducing an unacceptable amount of noise. Instead of the arithmetic mean, the point should be derived from the sines and cosines that, when summed, form the signal.

This is typically done by using an interpolation filter, which takes into account several points of history in order to provide the interpolated sample.

The figure below shows the effect of interpolating using the average of two adjacent points. The result looks smooth, with a little bit of roughness due to the interpolation. However, the real picture of the roughness comes from examining the spectrum: the averaging technique has generated a mirror-image signal at the far end of the spectrum. In the time domain, the interpolated signal "looks pretty good". But in the frequency domain, a new frequency component having significant amplitude has arisen.

Interpolating from adjacent points introduces more noise with higher frequency input signals, and with higher interpolation factors. For low frequency signals, the nearest neighbor average is closer to the "ideal" value, so the noise associated with lower frequency signals is lower.

Note the x-axis scales in Figure 78. In the time domain, the x-axis (sample count) is the same, but the signal has spread out because of the insertion of interpolated points. The x-axis in the interpolated result represents half the time of the original signal plot, since samples occur at twice the rate of the original signal.

In the frequency domain, the axis for the interpolated signal can accommodate twice the bandwidth. We expect this, since the higher sample rate can accommodate higher frequency signals.

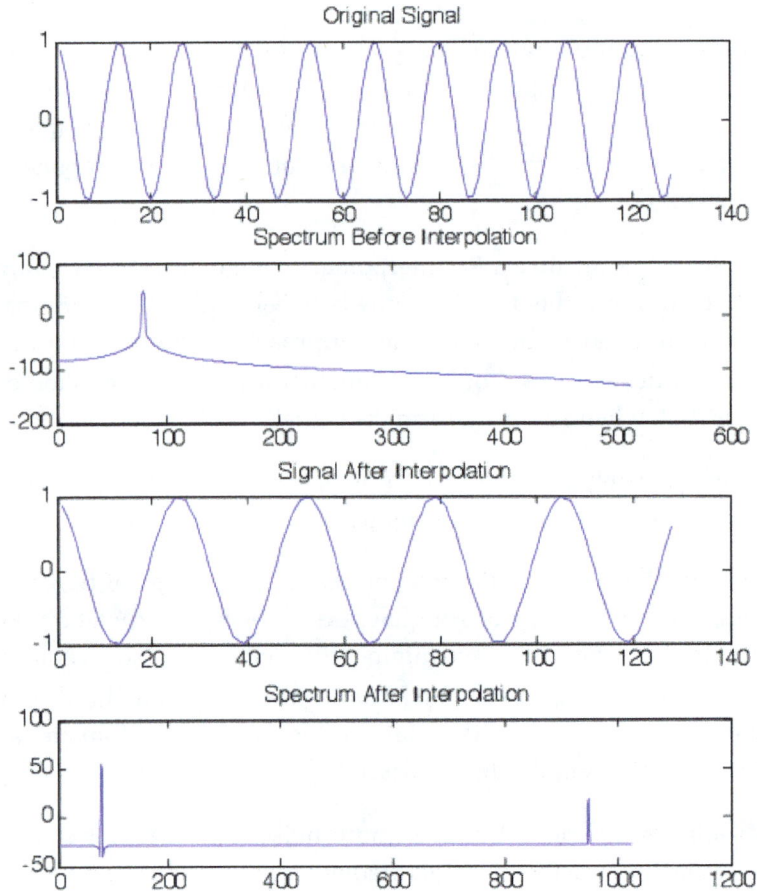

Figure 78. Interpolating using linear interpolation between points.

The Interpolation Process

The most straightforward way to interpolate a signal is to use the following process:

1. To interpolate by N, insert N-1 zero samples between each original sample. As unusual as this might sounds, this actually has a nice spectral quality which replicates the original spectrum N times.

2. Use a low-pass filter, operating at the new sample rate, to eliminate the unwanted copies of the original spectrum. The filter will have a passband equal to 1/Nth the full spectrum.

The quality of the low-pass filter determines the quality of the final result. The low-pass filter is much better than linear interpolation; the filter has the effect of curve-fitting the last several samples to generate each interpolation point.

Making Interpolation Faster

An interpolation filter processes a data stream consisting of N-1 samples having a value of 0 for every sample from the original input stream. The zero "filler" samples need not be multiplied or accumulated within the filter; this improves performance significantly.

An interpolation FIR filter need only operate on K/N samples for every output sample, where K is the number of filter taps. But, for every output sample, the subset of taps differs. As the input samples slide through the history buffer in Figure 79, only K/N taps will correspond with samples from the input stream.

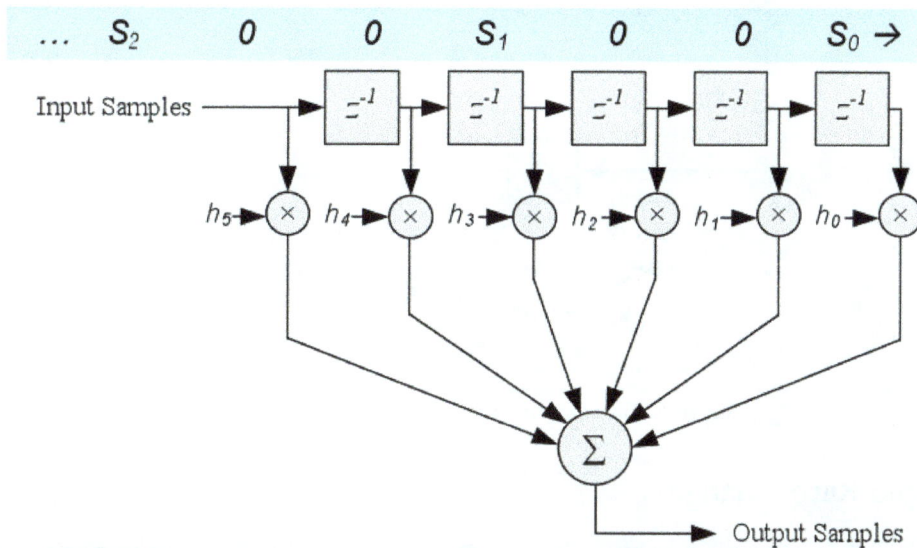

Figure 79. Interpolation Filter need not operate on 0-valued terms.

The interpolation by N filter can be restructured into N filters, each having K/N taps. The filters all receive the same data stream. However, each filter has a subset of the taps, separated by a stride of K. The output selector operates at N times the input sample rate, cycling through the filters in succession to select output points.

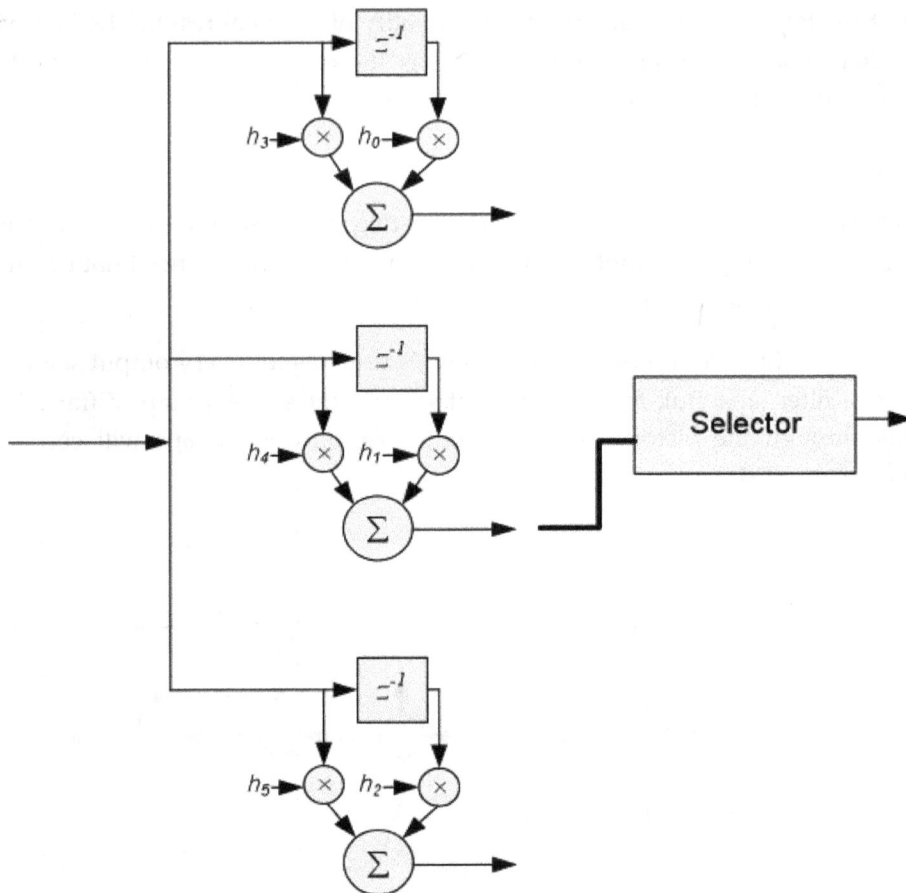

Figure 80. Restructured interpolation filter.

10.5 Sample Rate Changes

Decimation reduces a sample rate by an integer multiple; interpolation increases a sample rate by an integer multiple. What if we want to change a sample rate to a non-integer value? The answer is to decimate by M, then interpolate by N. (Or interpolate by M, then decimate by N). The filters can be combined into a single, more efficient structure which forms a matrix. This is called a *polyphase filter*.

10.6 Sample Rate Mismatch

When working with digitized audio signals that are played back in real-time, one soon discovers the following problem:

- The 8, 16, 22, or 44KHz sample rate at which the signal is digitized isn't exactly 8, 16, 22, or 44KHz; it is slightly off;

- The 8, 16, 22, or 44KHz clock used to convert the signal from digital back to analog is also slightly off, perhaps in a different direction.

- In a real-time system, where the source is connected to the destination through some path, these differences in clocks result in either an extra sample every several seconds, or a dropped sample every several seconds. (In a system where the data is stored and then played back, the different sample rates will just speed up or slow down the played back signal by a fraction of a percent.)

- The error is not constant; it is often varies as a function of temperature and component selections within the clock sources used within the system.

This raises the question: can't we just drop the extra sample when it occurs? Or, if we fall short, can't the system just repeat a sample?

The Intuitive (wrong) Answer

Examine the signal below. It is smooth, well-behaved, with a well-defined spectral peak. Dropping a single sample would hardly be noticed in the time domain; it would seem safe to do so.

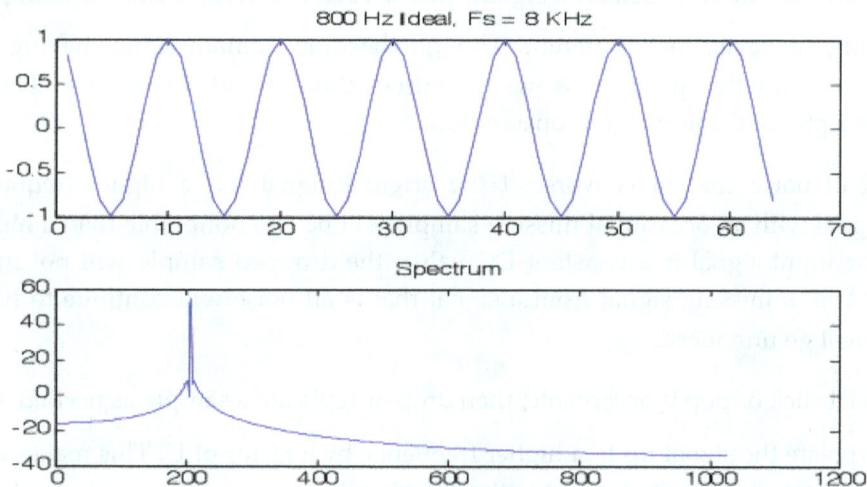

Figure 81. A well-behaved signal before dropping a sample.

163

The figure below show the same signal, superimposed with a version of the signal where a sample was dropped. In the time domain, the original signal (in green) looks very much like the signal with the missing sample (blue). The missing sample occurs at about sample 33.

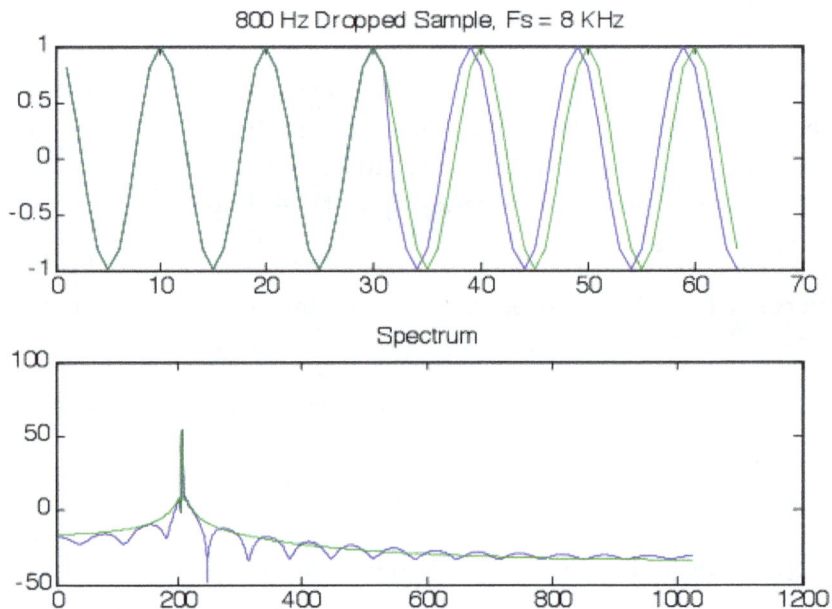

Figure 82. The well-behaved signal, with a version having a missing sample.

The spectrums, however, look different. Though the time domain signal having the missing sample looks smooth, the spectrum shows the introduction of wide-band noise. This will sound like a click or a pop in the midst of a constant tone.

The amount of noise content is worse if the original signal has a higher frequency. A low-frequency signal with an occasional missing sample will be less noticeable than a high-frequency sample. If the input signal is a constant DC value, the dropped sample will not introduce any noise at all. And, a missing signal from a signal that is all noise will continue to be all noise; a click or pop will go unnoticed.

If an occasional click or pop is acceptable, then drop or replicate a sample as needed. Otherwise:

- Interpolate the signal up to a higher frequency by a factor of U. This makes all frequency components lower with respect to the sample rate.

- At the new sample rate, instead of dropping or inserting one sample for every k samples, you must insert or drop U samples for every kU samples out of the interpolator.

- Reduce the sample rate back down: low-pass filter and decimate by U.

164

Exercises

1. Linear interpolation is the process whereby data between two points is estimated by assuming that the two points form a straight line; the estimated point is then proportional to the distance between the points on the X-axis and the slope of the line connecting the points. Since linear interpolation is computationally efficient, why use filtering to accomplish interpolation in the signal processing environment?

2. A system requires that the frequency components of a set of 200 samples be computed. An engineer proposes that the samples be padded out to a length of 256, so that a radix-2 or radix-4 FFT can be performed. What effect does zero padding the samples have on the magnitude of the FFT?

3. A certain system samples a sensor to produce a sample stream at 250K samples per second. The information of interest lies between 100 KHz and 120 KHz. Propose various approaches for operating on the sample stream that eliminates any signals outside of the 100 KHz to 120 KHz region.

4. Does decimation change the signal to noise ratio? If so, how?

References

[1] Harris, Fred, *Musical Applications of Microprocessors*, Hayden Book Company, a division of Hayden Publishing Company, Inc., Hasbrouck Heights, New Jersey/Berkeley, California, 1985, pp. 14-15.

[2] Lyons, Richard G.

Practical Applications

11

This section presents some approaches for common, practical DSP building blocks used for audio and RF processing.

11.1 Automatic Gain Control

Automatic Gain Control (AGC) originated in analog receivers as a way to automatically adjust the amplification stages within a receiver. AGC was essential for analog television receivers: the power difference between different stations can be large, based on distance, direction, and different transmitter elevations and radiated power. Within analog television receivers, the signal level determines the contrast: weaker signals show lower contrast. AGC allowed viewers to switch from channel to channel without having to adjust contrast.

AGC extends the dynamic range of a sensor. If a receiver has a 10 bit ADC, then the ADC provides a dynamic range of approximately 60 dB. If, however, the ADC is preceded by an adjustable gain stage providing 36 dB, the gain can be increased when the input signal (into the ADC) is weak, and decreased when it is strong, providing an overall dynamic range of 96 dB. The noise due to quantization will be limited to 60 dB.

Within a digital system, if the input sensor must be capable of collecting samples on a wide range of amplitudes and frequencies, the analog signal level into the ADC should be closely matched to the full ADC range. If the input signal exceeds this, clipping will occur, which introduces harmonics and destroys the input signal. If the input signal level is too low, then low-level input signal components may be lost due to quantization noise.

An approach to extending the dynamic range of a system is to precede the ADC with a variable attenuator. The attenuator is adjusted based feedback from a later processing stage. An AGC algorithm determines the "best setting" for the attenuator, providing near full-range without clipping.

A "best setting" should allow for some "headroom": that is, the attenuator should be adjusted so that the signal level into the ADC is the full scale value minus 6 dB or so. This allows a rapidly changing signal to occasionally exceed the "best" level without clipping.

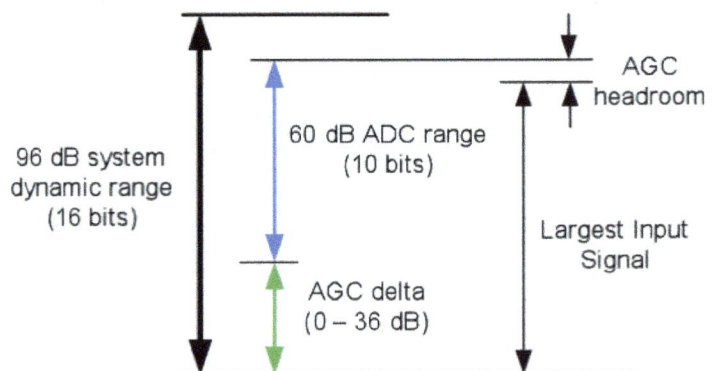

Figure 83. AGC Extends System Dynamic Range.

Attack Time

Attack time is the amount of time an AGC algorithm takes to respond to a large increase in signal level. If an ADC signals an over range condition, or if a sample value from the ADC is near the maximum value, the algorithm should reduce the signal level into the ADC rapidly, under the assumption that the signal level may continue to increase.

Decay Time

Decay time is the amount of time the AGC algorithm takes to respond to a sudden decrease in input signal level. If a strong signal suddenly disappears from the input, the AGC algorithm should increase the signal level into the ADC.

Many systems employ a fast attack, and a slow decay. The slow decay better accommodates the possibility that the strong signal will recur a short time after disappearing.

An AGC system must compute the average signal level into the ADC. This is done by taking the absolute value of all samples, and low-pass filtering the result. The low-pass filter will introduce a time delay proportional to the filter length and sample rate. The output of this filter is used to determine the value provided to the attenuator.

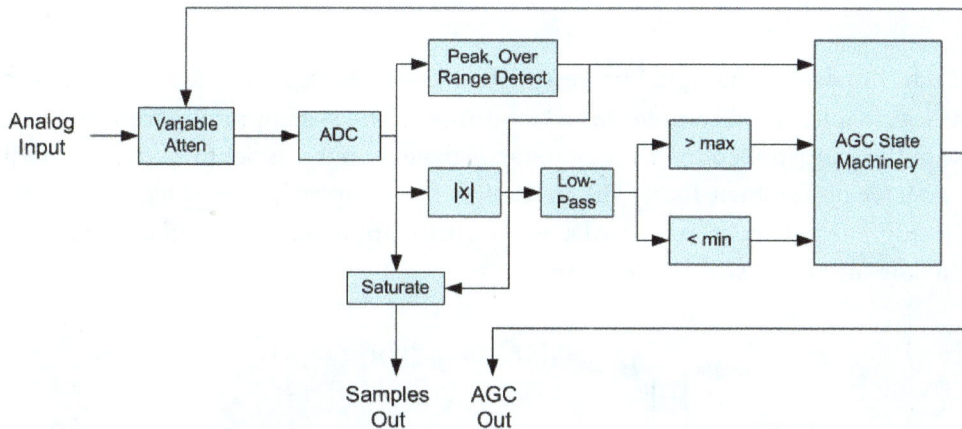

Figure 84. This AGC implementation adjusts the signal level into the ADC.

The AGC algorithm operates in one of three states: in the attack state, the filter output indicates a high signal level is present, and the AGC should increase attenuation. In the decay state, the filter output is slow, and the algorithm should decrease attenuation. If the average signal level falls within an acceptable middle zone, the AGC algorithm should not adjust the attenuator.

The low-pass filter must have a cutoff frequency that is well below the normal variations that input signals exhibit. For example, if input signals are amplitude modulated voice signals having a bandwidth of 100 to 4000 Hz, the AGC low pass filter should have a cutoff frequency of less than 100 Hz. Otherwise, the AGC will compensate for low frequencies within the modulated signal.

Programmable attenuators often provide discrete steps of 1 dB or ½ dB. A step adjustment to a signal stream, especially a large step adjustment, will introduce harmonics. AGC algorithms usually apply a "smoothed" set of small transitions to the input signal over time, avoiding large jumps in the signal level.

For ultra-fast attack, an AGC algorithm can look at the absolute value of the unfiltered samples, and also the over range indicator from the ADC. This severe case can be used to rapidly increase attenuation and avoid clipping when a very strong input signal suddenly appears. Of course this generates a large, rapid jump. Such a jump is probably less severe than clipping that would otherwise occur.

AGC Compensation

AGC generates an attenuator setting, which can be made available to subsequent signal processing stages. For example, subsequent filtering and demodulation blocks may wish to evaluate the signal level in order to determine if the input signal level was strong enough to warrant further processing.

One approach is to design the signal processing to accommodate a sample size that is wider than the ADC (for example, 16 bit samples for a 10 bit ADC). The system multiplies input samples by the inverse of the gain applied by the attenuator. If the attenuator is set to 24 dB, (i.e., reduces the signal by a factor of 16), then the FPGA multiplies the samples by 16. If a system is subject to technology upgrades such as wider ADCs, then this approach allows the current design to anticipate a more accurate ADC in the future.

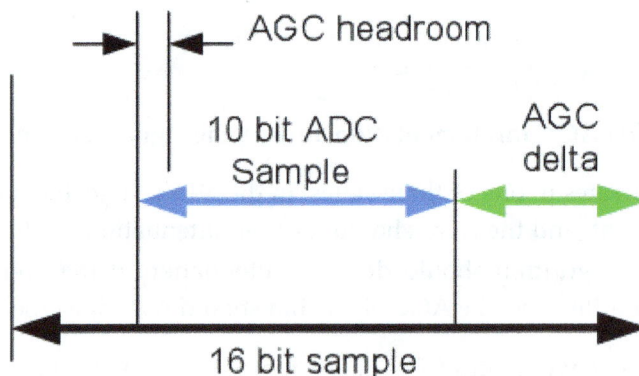

Figure 85. AGC Compensation aligns sample within a wider precision field.

AGC is essentially a low-pass IIR filter applied to the magnitude of the signal level. The feedback loop has delays, and a natural resonant frequency determined by the closed loop response time. An input that repeatedly increases in signal level and then decreases at the same time the AGC is adjusting the signal level for the increase will cause the gain value to oscillate to the point where it reaches the maximum and minimum gain values. Such conditions can be mitigated by using two low-pass filters (one with lower delay and therefore a higher cutoff frequency), and having the controlling state machinery base its decision on the signal levels seen by both filters.

11.2 Automatic Volume Control

The AVC (sometimes known as Automatic Level Control, or ALC) algorithms are closely related to AGC, in that they monitor the absolute value of a signal level through a low-pass filter, and then adjust the level. AVC normally operates on audio signals (that is, after demodulation), whereas AGC normally operates on RF and IF signals.

AVC is a convenience which prevents a listener from having to adjust the volume control when retuning or when presented with a signal exhibiting large volume swings. An AVC stage would normally be followed by a user-adjustable gain stage for manual volume control.

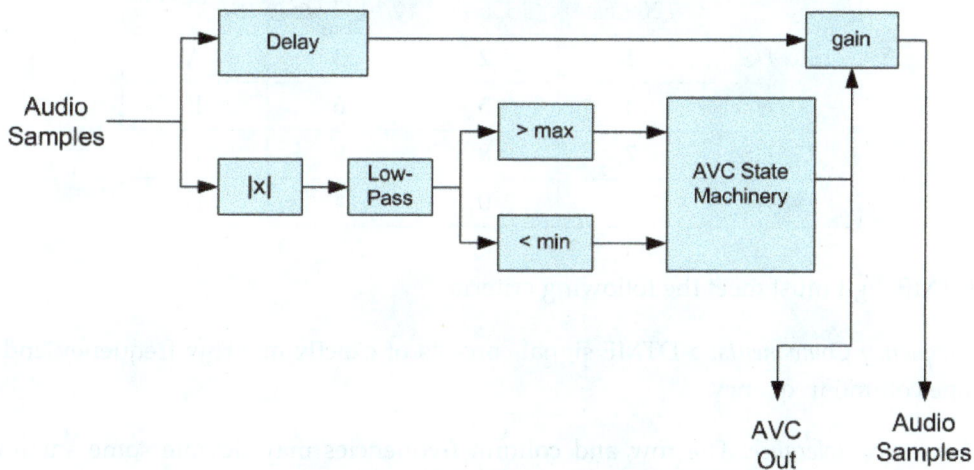

Figure 86. A typical automatic volume control implementation.

As with AGC, AVC can have different timing profiles for attack and decay. And, as with AVC, the low-pass filter employed must have a cutoff frequency that is lower than that of the lowest audio signal component that the AVC system must preserve.

The AVC diagram presented in Figure 86 uses feed-forward control to adjust the volume of the signal. The level computed by the low pass filter, and the gain determined by the state machine, are applied to the signal after the signal has been delayed. This differs from the AGC, which protects the current signal level into the ADC based on ADC samples acquired in the recent past, which is inherently a feed back process.

11.3 Squelch Control

Squelch is a process whereby low-level noise is replaced with silence. Squelch can be implemented by determining if the AGC and/or ALC levels are driving the level up from the noise floor. If so, a squelch algorithm replaces the audio signal with silence.

A more robust squelch can determine the degree of correlation between the input signal and noise, and suppress audio when the input is both low-level and correlates well with wideband noise.

11.4 DTMF Detection

Detecting frequencies within a sample stream is more complicated than finding "peaks" in the FFT. A DTMF digit consists of a single row frequency, added to a single column frequency, where the frequencies are chosen from the table below:

	1209 Hz	1336 Hz	1477 Hz	1633 Hz
697 Hz	1	2	3	A
770 Hz	4	5	6	B
852 Hz	7	8	9	C
941 Hz	*	0	#	D

A valid DTMF digit must meet the following criteria:

- *Frequency Components:* a DTMF signal consists of exactly one row frequency and exactly one column frequency.

- *Frequency Tolerance:* The row and column frequencies may deviate some small amount from the exact frequencies. Some DTMF sources will be RC oscillators, which will not be as precise as digitally synthesized sources.

- *Signal Level:* The row and column frequencies must be significant when compared with all other energy in the signal. They must also exceed a minimum threshold.

- *Relative Level:* The row and column frequencies may have different power levels, due to the frequency response of the communication channel and/or different RC oscillator output levels. The relative level is called the "twist," and is measured in dB. The twist for valid signals must not exceed a maximum value.

- *Duration:* A DTMF signal must last for a minimum amount of time (say, 50 mSec or longer), to be considered valid.

- *Dropouts:* An interruption in a DTMF signal of, say, 15 mSec or less, should not cause the detector to report two digits. An interruption of 50 mSec or more should cause the detector to report two digits.

- *Concurrent Dial Tone:* A DTMF signal may be coincident with a dial tone. This occurs in North American telephone systems when the first digit is dialed: the dial tone is present until the first digit is detected; there is a slight time overlap when both the digit and dial tone are present.

The goal is to design a detector that meets the above criteria and uses minimum resources. An 8KHz sample rate is assumed.

Approaches

We divide the DTMF detector into two parts: a frequency analysis part, which detects and reports the DTMF symbol associated with a segment (say, 50 to 60 mSec of input data), and a state machine, which processes a stream of symbol, and reports the final symbol value. The state machine detects and tolerates dropouts, ensures that symbols have a minimum duration, and reports a single symbol for an arbitrary duration.

Goertzel Algorithm Approach

An approach based on Goertzel is efficient for a software implementation. The input signal is filtered to remove the dial tone. The result is examined using the Goertzel with four row frequencies. If exactly one of the outputs constitutes, say, more than 35% of the total signal energy, then the signal is a candidate for further analysis; otherwise no digit is present.

The next stage applies Goertzel four more times, tuned to the column frequencies, to the filtered signal. If exactly one of the column frequencies constitutes, say, more than 35% of the total signal energy, then the signal may contain the digit indicated by the row and column frequency. Otherwise, no digit is present.

Additional processing confirms that the ratio of row to column amplitudes are within the acceptable twist limits.

Final qualification consists of measuring the second harmonics at the selected row and column frequencies (two more applications of Goertzel). If the second harmonic content is below a rejection threshold, then no digit is present.

FFT Approach

The FFT approach uses a single FFT instead of multiple applications of Goertzel. Windowing is required; a Hamming window is appropriate.

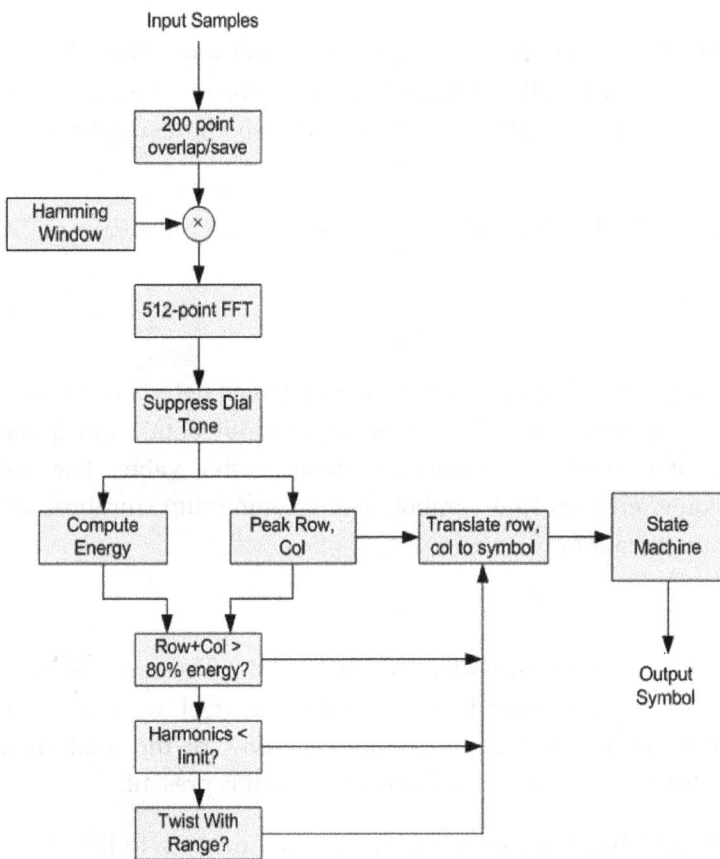

An FFT length of 512 corresponds to 64 mSec, and a bin width of 15.6 Hz. Some DTMF frequencies will span two adjacent bins; the sum of the energy in both bins should be used to determine the presence of a row or column frequency.

Because the start of a DTMF symbol is asynchronous with respect to the 512 sample frames, an overlap/save method precedes the FFT. An overlap of 256 points will examine 512 samples every 32 mSec; an overlap of 200 points will examine 512 samples every 25 milliseconds. A Hamming window suppresses high frequencies introduced by jump discontinuities.

Figure 87. An FFT-based DTMF Detector.

Dial tone suppression consists of zeroing out bins corresponding to 350, 440 Hz. Peak row and column search examines those bins representing valid DTMF tones, and reports the row and column having the maximum energy. Since some tones span adjacent bins, this function must combine adjacent bins.

The peaks are then qualified: the sum of the row and column energy must account for at least 80% of the total energy after dial tone removal; the second harmonic level must be below threshold and the ratio of row to column energy (twist) must be within range.

If all parameters are valid, the translation block uses the row and column indices to generate a valid DTMF symbol; otherwise it generates a "null" symbol.

The state machine requires a minimum of two matching symbols in a row (50 mSec) before deciding that a symbol is valid. Once decided, the state machine ignores single dropouts, and waits for the end of the symbol, indicated by a change in symbol for at least two 25 mSec time periods.

Some parameters (the 80% energy, the second harmonic reject level) can be further tuned to further refine the detector.

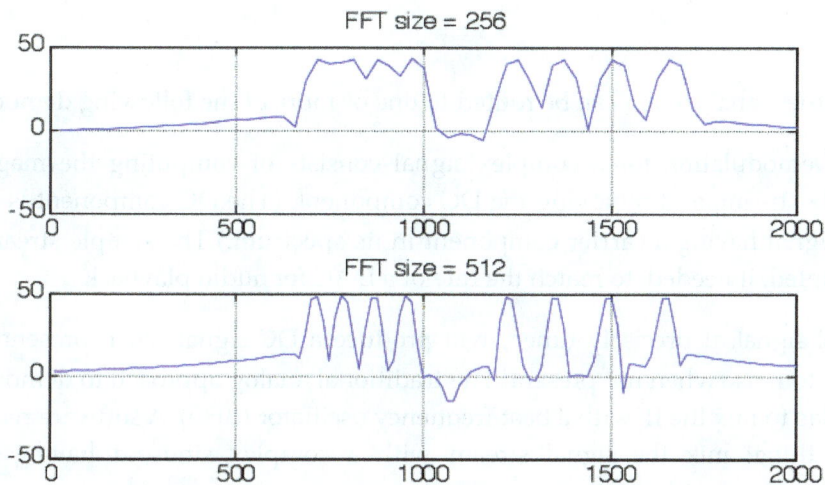

Figure 88. DTMF Frequencies analyzed with FFT sizes 256, 512.

11.5 Software Radios

Given the ability to downconvert, filter, and demodulate RF signals, one can construct a receiver that performs almost all of the signal processing using an FPGA or digital signal processor. The advantage of such a receiver is that it can be reprogrammed for new modulation types, different filters, and/or different encoding methods.

The front end of such a receiver often consists of an RF analog downconverter, followed by a high speed ADC. The downconverter provides a wideband intermediate frequency signal to the ADC. Firmware or software is then responsible for tuning specific signals within the wideband IF, demodulating, performing any decoding, and providing the results in a form suitable for consumption by a computer and/or a human.

The Gnu Radio Universal Software Radio Peripheral (USRP) is a representative device. A wide variety of open-source software and firmware is available for the device.

Tuning

The data presented to the ADC is normally a wide-band signal. Within this wide-band signal are potentially several narrower-band signals.

Tuning an individual signal is accomplished by mixing the ADC samples with a complex sinusoid having a frequency equal to the negative tuning frequency, and low-pass filtering the result. Normally this is followed by decimation, since the desired signal often has a much narrower bandwidth than the IF.

Demodulation

Once filtered, the signal stream can be routed to one or more of the following demodulators:

- AM Demodulation for a complex signal consists of computing the magnitude of the sample stream, and removing the DC component. (The DC component is a result of the AM signal having a carrier component in its spectrum.) The sample stream can then be resampled, if needed, to match the rate of a DAC for audio playback.

- A CW signal, if precisely tuned, will produce a DC signal when present (silence), and revert to noise when not present. The traditional analog approach to demodulating a CW signal is to mix the IF with a beat frequency oscillator (BFO). A software receiver does the same thing: mix the signal stream with a complex sinusoid having an adjustable frequency within the audio range.

- Single sideband reception also requires mixing with a BFO. This shifts the audio spectrum to the left or right. Typically the software receiver would be tuned to the center of the SSB signal, so that the low pass filter can eliminate noise above and below the signal. The resulting signal must then be shifted to the left or right so that 0 Hz occurs in the proper spectral location of the demodulated output. BFO mixing accomplishes this. Lower sidebands, which are spectrally reversed, must be mixed with a positive BFO frequency, resulting in an audio signal having positive frequency components.

- FM and PM signals can be limited (that is, clipped and filtered to remove AM), or, more directly, the sample stream can be converted from rectangular coordinates to polar coordinates, and the radius (magnitude) can be ignored. Then, for FM demodulation, the difference in phase from sample to sample (the first order difference, analogous to the derivative), should be computed. A de-emphasis filter is then applied for FM signals: this is a low-pass filter having a specific response curve. The de-emphasis filter compensates for pre-emphasis performed within the transmitter.

- FSK modem signals can be demodulated using the approach in the next section.

- PSK and QAM signals require "training" pulses that provide the receiver with a phase and amplitude reference. Once the receiver locks onto the reference, the receiver can determine the value of each symbol.

11.6 FSK Demodulator

The Bell 202 Modem uses frequency shift keying to pass data at 1200 bits per second. The modem uses a tone of 1200 Hz to represent a 1, and 2200 Hz to represent a 1.

The block diagram for a practical demodulator is shown below. Though designed for Bell 202, the parameters can be changed to accommodate any FSK signal. The input signal is center-tuned to -1700 Hz. The resulting complex signal has a frequency of -500 Hz (for a 1), and +500 Hz (for a 0), assuming a modem signal is present.

The 8K sample rate is not a convenient multiple of the baud rate, so the demodulator interpolates the signal by a factor of 3, equivalent to 20 samples per data bit.

The sample stream is then normalized and the phase change from sample to sample is estimated. At 24K samples per second, the phase change value for ±500 Hz will be ±0.1309 radians. Higher frequencies will be eliminated by the low-pass filter; lower frequencies will provide a lower value. The signal level detector is a decision operator examines the phase change values for the last several samples to ensure $|x|$ is approximately 0.1309 for the last several samples. Small dropouts are acceptable; these will occur where the modem signal transitions from 0 to 1, or from 1 to 0.

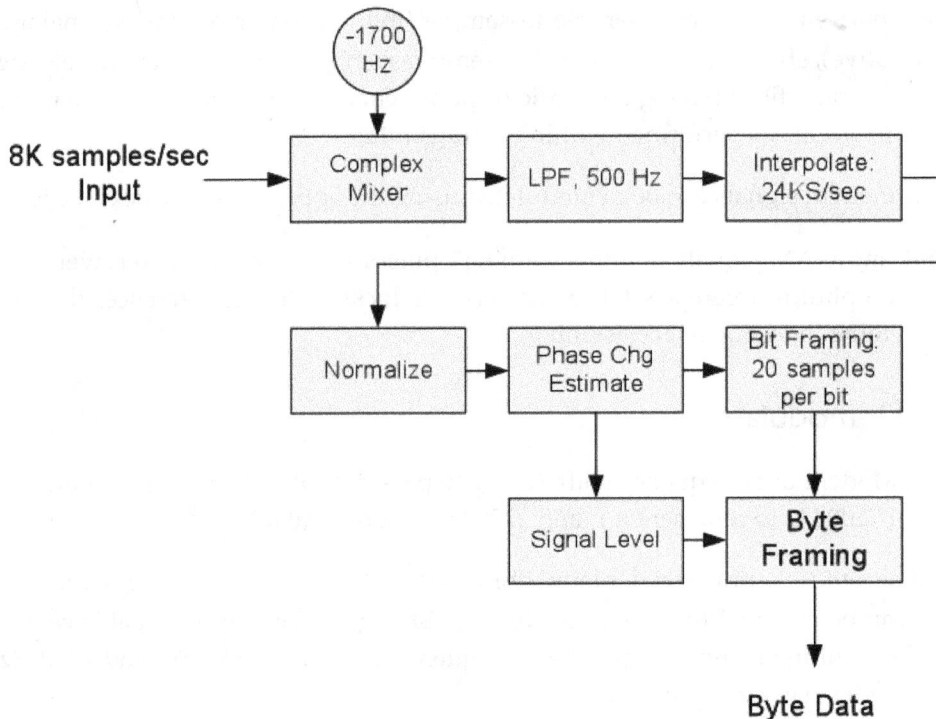

Figure 89. An FSK Demodulator.

The bit framing finds changes in sign in the sample stream; these correspond to data changes from 0 to 1 and from 1 to 0. Once a transition is located, the bit framing counts off 10 bits (to the center of the bit) before generating a bit value to the byte framing block.

Not all data streams will have transitions: the bit pattern 00001111 will have a single transition for eight bits. The bit framer therefore generates a new output after every 10+20n bit times for any integer n. The modem stream normally contains a set of 1-0 transitions before the data starts, allowing the bit framing block to synchronize with the data.

The byte framing block examines the sequence of blocks, gated by the signal level detector. Byte framing can vary with serial data modes; an asynchronous byte framer will look for a start bit (a transition from 1 to 0), followed by a programmable number of data bits, an optional parity bit, and at least one stop bit.

Dot Product Applications

Appendix A - Dot Product Applications

Definition: The *dot product* of two vectors is the sum the product of corresponding points. Often, if the second vector is complex, the conjugate value of the second parameter is used.

$$X \cdot Y = \sum (x_i \, y_i)$$

RMS: The root-mean-square of a vector is a measure of the energy. The RMS for a vector of length *len* is computed using the dot product as follows:

$$RMS(X) = \sqrt{[(X \cdot X) / len]}$$

RMS Error: The RMS distance (error) between two vectors is computed by subtracting corresponding elements, producing a new vector, and then computing the RMS of the resulting vector:

$$\text{Error Vector } E_i = (x_i \, y_i)$$

$$RMS \text{ Error } (X, Y) = \sqrt{[(E \cdot E) / len]}$$

Variance: The variance of a vector is the dot product of the a vector representing the distance from the mean with itself, divided by the vector length:

$$var(X) \quad = [(X - \mu) \cdot (X - \mu) / len]$$

Standard Deviation: This is the RMS distance about the mean. It is also the square root of the variance:

$$Stddev(X) \quad = RMS(X - \mu)$$

Correlation: The correlation between two vectors is a measure of how well increases and decreases in one vector correspond to increases and decreases in the other vector. Large, positive values indicate good correlation; large negative values indicate good inverse correlation; small values indicate poor correlation. The correlation is the dot product between two vectors:

$$\text{correlation} = \sum (x_i \, y_i)$$

The correlation is often computed by computing the dot product of an input vector with a pattern, then sliding the input vector over to the next sample, and recomputing, producing a correlation vector:

$$c_k = \sum_{k=0}^{len-1} (x_{i+k} \, y_i)$$

Correlation Coefficient: This is single value between -1 and 1 indicating the degree to which two vectors match. A value of 1 is a perfect correlation; a value of -1 is a perfect inverse correlation, values close to 0 indicate poor or no correlation. The vectors are normalized, guaranteeing that the result is between -1 and 1:

$$r \; = \; \frac{1}{N} \times \frac{(X - \mu_x) \cdot (Y - \mu_y)}{\sigma_x \sigma_y}$$

Convolution is similar to correlation, except the order of the second vector is reversed:

$$c_k = \sum_{i=0}^{len-1} (x_{i+k} \, y_{len-i-1})$$

FIR Filter: A FIR filter is the convolution of an input signal with a vector, known as the impulse response of the filter. The impulse response is vector is often called *h*.

$$c_k = \sum_{i=0}^{len-1} (x_{i+k} \, h_{len-i-1})$$

Goertzel Algorithm is the dot product of an input vector with a complex sinusoid of frequency *f* sampled at a rate of Fs. The result is a single complex scalar value. The result indicates the magnitude and phase of the component of frequency *f* within the input signal.

$$G_\omega = \sum_{k=0}^{len-1} (x_k \, e^{-2\pi jkf/Fs})$$

DFT: The discrete Fourier transform of a vector having length *n* is a set of Goertzel algorithms computer for frequencies 0 to *n-1*.

Glossary of DSP Arithmetical Terms

Appendix B - Glossary of DSP Arithmetical Terms

ADC Bias

Some Analog to Digital converters provide a biased result. For example, the ADC08D1520 8-bit converter converts the differential input voltage to an unsigned 8-bit binary value between 0 (the most negative voltage) and 255 (the most positive voltage). Using this scheme, 0 volts has a value of 128:

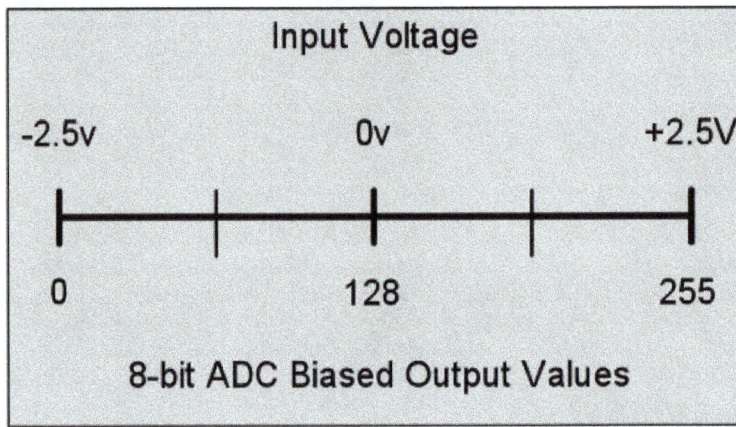

A more desirable representation is to represent the most negative voltage by -128, 0v by 0, and the most positive voltage by 127.

The output from the ADC is biased; it is 128 higher than the desired value.

The biased value can be converted to an unbiased value by subtracting the bias, 128.

Shortcut: inverting the MSB of an 8-bit value is equivalent to subtracting 128.

As proof, note that -128 is binary 1000_0000. Adding this value to an 8-bit value will always complement the state of the high-order bit of the 8-bit value.

Alignment

When adding, subtracting, or comparing two values having a different number of bits to the right of the radix point, the numbers must first be aligned:

```
0110.10 + 111.001:

0110.100          padded with one trailing 0
1111.001          sign extended
-----------
0101.101
```

Remember that when aligning signed values, if a number needs to be padded on the left, then it must be sign-extended, not simply zero padded.

Bit Growth

When two 8 bit values are added, there is a possibility that 9 bits may be required to represent the correct result. Consider adding the signed values +127 and +5. The result is 132. However, since the maximum signed value representable within 8 bits is 127, 9 bits are required to represent the result.

When adding or subtracting, one more significant bit is required to represent the result.

Bit growth also occurs with multiplication. When multiplying two 8 bit values, the result may require 16 bits for accurate representation. 127 * 127 = 16129; this value can be represented using a signed, 15 bit integer. However, -128*-128 = 16384; this value requires 16 bits for signed representation.

When multiplying two values, one m bits in length, the other n bits in length, the result is m+n-1 bits in length.

Bit growth can be mitigated by one of the following strategies:

- High order bits can be truncated (see Modulo arithmetic). Generally this is not the best choice.

- The result can be saturated before truncating high bits. This is generally a better solution than truncation.

- Low order bits can be truncated.

- The result can be rounded before truncating low order bits. See rounding.

High-order truncation or saturation is often used when operating on integers. Low order truncation or rounding is often used when operation on fractional values. Both approaches (rounding and saturation) may be required in some cases.

Decibels (dB)

A decibel is a ratio between to quantities having the same units of measure. The decibel is based on the log base 10 of the quantities, so a small number of decibels may represent a large power.

Systems engineers found that within systems (and nature), some performance ratios were quite large. For example, the ratio of the softest detectable whisper to a sound level that causes pain is 1:1,000,000 or so. The ratio of the output power at the antenna of a radio station to the input power at a receiving antenna 10 miles away may be 1,000,000,000 to 1.

Rather than express these ratios as large values, the decibel is the preferred unit of measure. For power ratios,

$$dB = 10 \cdot \log_{10}(P_{out}/P_{in})$$

For voltage or current ratios and a fixed impedance, doubling the voltage (or current) also doubles the current (or voltage), resulting in 4x the power. Therefore, the power gain is equal to 20 times the log of the voltage ratio:

$$dB = 20 \cdot \log_{10}(V_{out}/V_{in})$$

When viewing the output of a signal generator into a fixed impedance on an oscilloscope, increasing the output level by 6dB doubles the voltage (and current), increasing the output power by a factor of 4.

Dot Product

The dot product between two vectors is the sum of the product of corresponding elements within the vectors:

$$X \cdot Y = \sum (x_i y_i)$$

The dot product is a scalar value. The value will be complex if either input vector is complex. When the y vector is complex, the conjugate value is often used within the dot product. If x and y are the same vector, and if x is complex, then $x_i \times conj(x_i)$ is real for all vector elements, and the result is real.

If the dot product of two vectors is zero, the vectors are said to be *orthogonal*.

The dot product forms the basis for these other signal processing functions:

- The RMS value of a vector is computed by taking the dot product of a vector with itself. This value is then divided by the length (to compute the mean), and the square root of the value is taken.

- The variance of a vector can be computed by subtracting the mean, and computing the dot product of the resulting vector. Dividing this by the vector length gives the variance.

- The dot product of a vector with a complex sinusoid is the Goertzel algorithm.

- The dot product of a vector with a pattern is the correlation.

- The dot product of a vector with a pattern is the convolution, and is the basis for FIR filters.

Dynamic Range

Within a number system, the dynamic range is the ratio of the largest representable value to the smallest representable value. For 16 bit two's complement values, this is 32767/1.

The dynamic range of a physical system is the ratio of the largest signal that can be successfully processed to the smallest signal that can be successfully processed.

They dynamic range is normally expressed in dB.

For human hearing, the softest perceivable sound is 10 dB, and the threshold of pain is 130 dB. The dynamic range for human hearing is 130 dB – 10 dB = 120 dB. To convert this to a linear ratio,

$$\text{Linear ratio} = 10^{(120 \text{ dB} / 20)} = 10^6 = 1,000,000$$

Therefore, the softest perceivable sound is about a millionth as loud as a sound level that causes pain.

An analog to digital converter provides a dynamic range, in dB of power, of approximately 6dB × the number of conversion bits. An 8 bit ADC provides a dynamic range of about 48 dB, a 12 bit ADC will provide about 72 dB of dynamic range.

CD players provide a dynamic range of about 96 dB, suggesting that CDs use 16 bit data to represent audio signals.

FIR Filter

A *Finite Impulse Response* filter is a digital filter. Each output sample is computed from the weighted sum of the last N input samples. The weights are called filter taps, and determine the behavior of the filter.

Filter taps are normally computed using specialized tools or algorithms. It is possible to design a FIR filter with almost any frequency response, but there is often a tradeoff between the number of taps and the number/sharpness of "transition regions" within the frequency response.

The FIR filter is implemented by computing the dot product between a set of input samples, and the taps, in reverse order (i.e., the convolution of the input signal with the taps). This generates one single output sample. The input samples are then "slid over" one sample, and the process is repeated to generate the next output sample.

FIR filters are advantageous because they are always stable. Also, a FIR filter can have a linear phase response within the passband (critical for DSP applications like phase modulation).

Fixed Point Representation

In contrast to floating point representation, fixed point representation uses a specified number of bits to represent a value. An implied radix point position within the bit field never changes, so the value always has the same number of integer and fraction bits.

Fixed point representation may be used to represent integers, or fractional values, or numbers that have integer and fraction bits.

Nomenclature varies, but a 16 bit signed value that has 10 integer bits and 5 fraction bits might be called an "S10.5" value, indicating a sign bit, 10 bits to the left of the radix point, and 5 bits to the right. Similarly, an S3.4 value is a signed 8-bit value having 4 bits for the fraction portion.

Fixed point implementations must deal with bit growth, overflow, and rounding, whereas these features are often built in to all floating-point implementations.

Floating Point Representation

Floating point representation is a means by which a limited number of the most significant bits of very large or very small values are stored, along with an exponent value for scaling. This is analogous to scientific notation.

Floating point representations are generally IEEE-754 compliant. This specification supports the following features:

- Support for single, double and extended precision values
- Representations for 0, -0, Not a Number (NAN), -∞, and +∞
- Rounding and guard bits used during operations
- Denormalized (extremely small) numbers
- Exception processing

Not all features are useful in signal processing.

The most useful feature is the greatly expanded dynamic range provided by floating point representations.

A single precision floating point number has this form:

Sign   Biased Exponent   Mantissa

The sign bit is 1 for negative values, and 0 otherwise. Floating point employs sign-magnitude representation.

The biased exponent is an 8 bit unsigned value that is 128 higher than the "actual" exponent value.

The mantissa is 23 bits long for a single-precision value. Except in the case of + or – 0, and NAN values, and denormalized values, the mantissa has an "implied 1" as the most significant bit. This bit is never stored.

Because of the complexity of processing floating point numbers, FPGA algorithms typically employ integer or fixed point operations. For example, the following steps are needed to add two floating point values:

1. Examine operands to determine if either is NAN:

   a. If either is NAN, the result is NAN

2. Examine operands to see if either is ∞.

   a. If both are ∞ but signs differ, force result to NAN

   b. If both are ∞ with same signs, the result is the same as either input

3. Align both operands according to exponent values.

   a. For normalized values, provide a leading 1.

   b. For 0 or denormalized values, do not provide a leading 1.

4. Complement operands whose sign bit is set

5. Add the values

6. If result < 0, complement and indicate sign of result is negative.

7. Check for overflow: if overflow, adjust exponent by +1, shift data right 1

8. Normalize: find the most significant nonzero bit in the result

    a. Zero detect: if the result is 0, force all bits (sign, exponent, mantissa) of the result to zero

    b. Otherwise, adjust resulting exponent by normalization bit count; shift the mantissa left or right as needed.

9. If the resulting exponent exceeds max, force result to $+\infty$ or $-\infty$.

10. If the resulting exponent is too low, represent the result by its denormalized value.

11. Set bits within the floating point condition code register to indicate that the result is > 0, < 0, or 0.

Fractional Representation

It is sometimes convenient to implement a digital signal processing system where all numbers are signed (or unsigned) fractional values, having no bits to the left of the radix point other than a possible sign bit. Thus, all sample values have decimal values between -1.0 and approximately +0.9999.

The chief advantage of this representation arises during multiplication.

When multiplying two integer values having bit widths of m and n, the result is m+n bits long. If both factors are small, the result will be in the low part of the m+n bit result. If both are large, the result will be in the high part of the result. It is not clear whether one should round and shift right and eliminate low bits in order to contain the bit growth, or truncate high bits and saturate.

When multiplying two fractional values having bit widths of m and n, the result is also m+n bits long. However, multiplication of two factors, each of which has magnitude of 1.0 or less, results in a product that is 1.0 or less. Therefore, the result can always be rounded back to m (or n, if desired) bits, providing a fractional result between -1.0 and 0.9999 that is ready to be used as an input into the next functional block.

FFTs and filters can be designed so that the maximum output is a fraction between -1.0 and 0.9999. Also, the results from an ADC can be treated as signed fractions between -1.0 (representing the negative full scale value) and 0.9999 (the positive full scale value).

Gain

A system that amplifies or increases the level of a signal provides gain.

Gain is the ratio of the input signal level to the output signal level. While gain can be expressed linearly, it is normally expressed logarithmically, in dB.

An audio amplifier that amplifies a 1 milliwatt signal to a 10 watt power level provides a gain of $(10/0.001) = 10,000$. Expressed in dB, this is

$$\text{Gain} = 10 \cdot \log_{10}(\text{Power}_{out}/\text{Power}_{in})$$

$$= 10 \cdot \log_{10}(10,000)$$

$$= 10 \cdot 4$$

$$= 40 \text{ dB}.$$

The opposite of gain is *loss*. Loss can also be expressed as a ratio or dB. Using dB, a loss of 30 dB is equivalent to a gain of -30 dB.

When systems are cascaded together, the gains, expressed as ratios, are multiplied. When using dB, the gains are added, since adding in the log domain is equivalent to multiplying linear values.

For example, a preamplifier increases the signal level by a factor of 10. This is fed into a power amplified that increases the signal level by a factor of 50. The resulting signal passes through a long cable that loses half of the signal level, providing a gain of ½. The resulting output signal has a level computed as follows:

$$\text{Output} = \text{preamp} \times \text{power amp} \times \text{cable loss}$$

$$= 10 \times 50 \times \tfrac{1}{2}$$

$$= 250$$

Expressed as dB, the preamplifier provides 10 dB of gain, the power amplifier provides $\log_{10}(50)$ = 16.9 dB of gain, and the cable loss provides -3 dB of gain:

$$\text{Output(dB)} = \text{preamp} + \text{power amp} + \text{cable loss}$$
$$= 10 + 16.9 - 3$$
$$= 23.9 \text{ dB of gain}$$

Converting back to ratios to check,

$$\text{Output ratio} = 10^{(23.9/10)} = 245, \text{ which is sufficiently close.}$$

IIR Filter

An Infinite Impulse Response filter is a digital filter. Unlike a FIR filter, an IIR includes feedback terms, that cause the next output to depend upon the next M input samples plus the weighted sum of the last N output samples. Therefore, filter's output samples may not settle at zero if the filter is presented with an impulse input.

When implemented using fixed point, the results may converge at zero after a finite number of terms. However, the filter is still an IIR filter, because it contains feedback paths.

IIR filters generally require less taps, or coefficients, to achieve filters with similar behavior to FIR filters. However, IIR filters have nonlinear phase response, near and at transition regions. IIR filters may also be unstable, in that certain inputs may result in outputs that grow without limits. Also, quantization of both the inputs and filter coefficients may result in unstable behavior.

Magnitude

The magnitude of a real number is its absolute value.

The magnitude of a complex number is the distance, in the complex plane, between the number, and zero:

The magnitude of (3, -2) is sqrt $(3^2 + (-2)^2)$

$$= \text{sqrt } (9+4)$$
$$= \text{sqrt } (13)$$
$$= 3.6$$

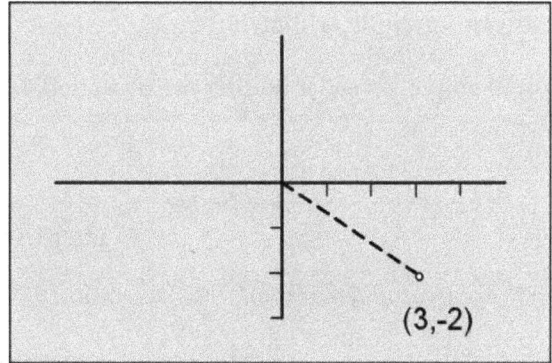

(3,-2)

Magnitude Squared

When dealing with complex values, it is sometimes easier and faster to use the magnitude squared value, rather than the magnitude. This avoids the square root computation (see Magnitude).

For example, when comparing two complex values to see which is closer to zero, one can compare the magnitude or the magnitude squared.

The magnitude squared of (3, -2)
is $(3^2 + (-2)^2) = 13$

The magnitude squared of a real number is simply the square of the number.

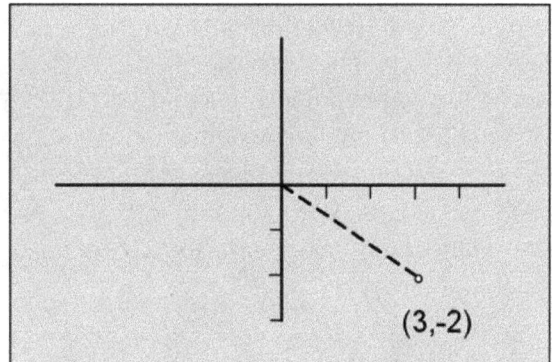

(3,-2)

Modulo Arithmetic

Modulo arithmetic (as opposed to saturation arithmetic) refers to adders, subtractors, and multipliers that, when overflow occurs, provide a result that is the remainder, or modulo, of the correct result, and a power of two.

A normal adder or subtractor, without additional logic to detect overflow and handle saturation, performs modulo arithmetic.

For example, consider adding the unsigned 4 bit values 11 and 13:

```
        1011        = 11
+       1101        = 13
---------------
       11000        = 8  after truncating the high bit
```

In this case, the result is (11 + 13) modulo 16 = 8.

Most DSP chips provide hardware that performs saturation arithmetic, whereas general purpose processors normally support only modulo arithmetic.

Modulo arithmetic is useful for circular buffers having a length which is a power of 2. If a buffer contains 64 entries, then the entry "after" buffer[63] is buffer[0]. This is also useful for sin and cos lookup tables having 2^n entries.

The graph below shows an input vector represented using 8 bit two's complement data. When we add 45 to each data point, some of the data points exceed the maximum representable positive value (127). In modulo arithmetic, these values "wrap" back to negative values. With saturation, the values are "clipped" to a maximum of 127.

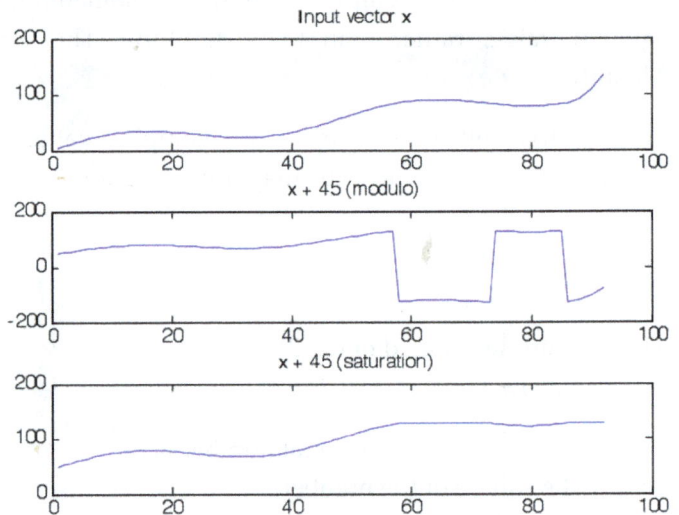

Figure 90. Saturation versus modulo arithmetic.

Overflow

Overflow occurs when the result of an operation cannot fit within the number of bits allocated for the result.

The following operations can cause overflow:

- Addition and subtraction
- Multiplication
- Truncating a value (discarding high bits)
- Negating the largest negative value can result in overflow: for 16 bit signed integers, the "largest" negative value is 0x8000, or -32768. The positive value 32768 cannot be represented in a signed 16 bit field, since the maximum positive value is 32767.
- Rounding a value can cause overflow if a "round up" occurs on a value that is already a maximum value: (e.g., rounding the signed 16 bit value 0x7fff.ff to the nearest integer would result in 0x8000, which is a negative value).

Overflow can be handled using one of the following methods:

1. Truncation: ignore the high bits of the result. This is also referred to as modular arithmetic.

2. Saturation: detect the overflow, and replace the erroneous result with the largest value that can be represented within the result field.

3. Provide a wider output field (that is, avoid overflow by accommodating the bit growth of the result). For signed addition or subtraction, the input values to the adder or subtractor should be sign-extended to the new width before performing the operation.

When adding two unsigned values, overflow can be detected by sensing the carry out bit. If this is ever nonzero, an overflow has occurred.

When adding or subtracting two signed values, overflow is detected by comparing the signs of the two input values: if the two input values have the same sign, and if the result has a different sign bit than either input, then overflow occurred. (Note that overflow can never occur when adding two signed values that have different signs.) Logically, signed addition overflow is detected using:

OFL := NOT (sign (operand1) XOR sign (operand2)) AND

sign (result) != sign (operand1)

Quantization Noise

When a signal is digitized, specific voltages are represented by fixed point binary values. An 8 bit ADC can provide one of 256 output values for any voltage input. If the 256 values represent a voltage level between 0 and 1 volt, then each integer "step" in the output value (sample) corresponds to 1/256th of a volt.

Because the input voltage will vary continuously with time, it will take on values that are "in between" values that can be represented by the ADC. In this case the ADC provides a sample value that is closest to the actual input voltage.

Because the sample value is slightly "off", a set of samples from an ADC does not perfectly represent the input signal.

If one feeds a pure sine wave into an ADC and examines the resulting spectrum, the expected result (a single, clean peak) will be accompanied by some amount of noise. This noise can be loosely characterized as random data that occurs because of the "quantization" of the input signal: forcing the samples to integer values.

There are two ways to reduce quantization noise:

- Use an ADC providing more output bits. Each bit will reduce the quantization noise by about 6 dB.

- Match the input signal range to the range of the ADC. If the ADC can process a signal that is one volt peak to peak, make sure that the input signal is amplified to a level that is approximately one volt peak to peak.

Radix Point

When using base 10, the decimal point separates the integer portion of a value from the fractional portion. For bases other than 10, the base is called the radix, and the point is called the radix point.

Round to Nearest

When truncating least significant bits, accuracy is enhanced by rounding to nearest before truncating.

As the name implies, rounding to nearest rounds a positive value up to the next integer value if the fractional part of the input value is greater than or equal to 0.5, so that 3.7 becomes 4, and 3.1

becomes 3. Negative values are rounded "down" to the "next" negative integer value if the fractional part is less than -0.5, so that -4.6 becomes -5, and -3.1 becomes -3.

> *Tip: Rather than testing a value to see if the fractional value is greater than or equal to ½, and conditionally selecting the next highest integer, rounding is most easily accomplished by adding ½ and then discarding all fraction bits.*

For example, 3.7 + 0.5 = 4.2; truncating the fractional part gives 4. This also works with negative values: -4.6 + 0.5 = -4.1; truncating the fractional part of the two's complement value the result is -5.

Rounding need not be limited to providing integer results. A number having 8 integer bits and 5 fraction bits can be rounded to 8 integer bits and 2 fraction bits using round to nearest. The implementation is more generally stated as follows:

> *To round off the low N bits, rounding to nearest, add $2^{(N-1)}$, and then discard the low N bits of the result.*

For example, to round off the low two bits of 0110110 (54) using round to nearest:

```
        0110111
+       0000010
-------------------
        0111001      Truncate low 2 bits
        1110         Result = 14
```

Since 54/4 = 13.5, 13.5 rounded up is 14, the correct result.

> Note that rounding can result in overflow. For example, rounding 011110 to nearest by two bits gives 011110 + 010 = 100000, which is a negative value. Rounding algorithms should therefore detect for overflow and employ saturation if needed.

Other forms of rounding include rounding towards zero, rounding towards +infinity, rounding towards –infinity. These other forms are used less frequently.

Saturation

Overflow can occur during addition, subtraction, negation, rounding, truncation, or other operations. When overflow occurs, the result can be truncated (yielding an incorrect answer that may be wildly different from the correct answer), or saturated (yielding the largest positive or negative result that can be represented within the resulting bits).

Saturation is often desirable; although it is numerically inaccurate, it is much closer to the actual result than a result obtained by truncating high order bits.

See *modulo arithmetic* for a graphical example of modulo versus saturation arithmetic. Note the huge jump discontinuities introduced in modulo arithmetic; these are not present in saturation arithmetic. If the data set is to be subsequently processed in the frequency domain, the modulo arithmetic model introduces significant high-frequency noise when overflow occurs; the saturation model introduces much less noise.

Operation	Expected Result	Truncated Result	Truncated Error	Saturation Result	Saturated Error
Signed 0100 + 0101	01001 = +9	1001 = -7	16	0111 = +7	2
signed 1001 + 1010	10011 = -13	0011 = +3	16	1000 = -8	5
unsigned 1001 + 1010	10011 = 19	0011 = +3	16	1111 = 15	4

When saturating the result of a signed operation, the expected sign of the correct result must be determined. If the expected result is positive, the saturated result must be forced to the most positive value: 0111…11. If the expected result is negative, the saturated result must be forced to the most negative value: 1000….000. When saturating an unsigned result, the result must be forced to all 1s.

Sign Extension

When adding or subtracting values having different bit widths, the shorter value must be sign-extended. Sign extension works as follows:

- When sign extending a negative value, pad the value on the left with leading 1 bits
- When sign extending a positive value, pad the value on the left with leading zeroes.
- When extending an unsigned value, pad the value on the left with leading zeroes.

Example: Add the four-bit signed value 0110 (decimal 6) to the three bit signed value 111 (decimal -1):

```
0110
1111     sign extended
------
0101     result = decimal 5
```

Sign-Magnitude Representation

Sign-magnitude representation of a numerical value uses the most significant bit of a binary number as a flag to indicate that the value is negative; the remaining bits are the "unsigned" value of the number. It is customary to set the most significant bit to indicate a negative value.

For example:

> 0101 = positive 5,

> 1101 = negative 5.

Though sign-magnitude is a little easier for humans to deal with, in practice it is used infrequently, for the following reasons:

- When adding, subtracting, or comparing, either operand having a high bit set must be two's complemented "on the fly"

- It is possible to represent both +0 and -0 using sign magnitude: for a four bit field, +0 = 0000 and -0 = 1000. When comparing values for equality, special logic must handle this case since we want -0 to equal +0.

For fixed point representation, two's complement representation is preferred over sign-magnitude representation.

Truncation

Truncation refers to discarding bits in order to reduce the number of bits required to represent a result. Truncation can occur on the left (i.e., most significant bits) or the right (i.e., least significant bits).

Truncation is often applied to the result of a multiply, since multiplication results in large bit growth.

Truncation on the left can result in overflow, and is often handled using saturation. When truncating a signed value, all discarded bits must have the same value as the sign bit of the result, otherwise overflow occurs. Consider the 9 bit signed values below, truncated to 5 bits by discarding the high 4 bits:

Truncated bits	Remaining bits	Overflow?
0000	10110	yes: input > 0, result < 0
1111	10010	no: input and result both < 0
0101	01001	yes: truncated bits not all 0's or 1's
1101	11100	yes: truncated bits not all 0's or 1's

When truncating an unsigned value, all discarded bits must be zero, otherwise an overflow occurs:

Truncated bits	Remaining bits	Overflow?
0000	10110	no: input > 0, result > 0
1111	10010	yes: discarded bits not 0
0101	01001	yes: discarded bits not 0

Truncation on the right is often handled using rounding. Note that rounding can result in overflow.

Two's Complement Representation

Negative binary numbers can be represented using either two's complement representation or sign-magnitude representation.

With two's complement representation, the negative of a number is computed using the following steps:

- Invert all of the bits,
- Add one to the result.

For example, using 4 bit fields, the value 5 is 0101. The two's complement value -5 is computed by inverting all bits (0101 -> 1010), and then adding 1 (result = 1011).

A two's complement value is negative if the high order bit is set.

Unlike sign-magnitude, two's complement values can be added directly, using a normal full-added, with no special preprocessing as is required using sign-magnitude. If the resulting value is negative, it is also a two's complement value. Also, when computing the difference between two's complement values A – B, rather than negating B, it can be inverted, and the +1 operation can be performed by setting the carry input to the full adder. Two's complement is therefore preferred because it results in simpler, faster hardware implementations.

Unsigned Arithmetic

If a variable can take on only non-negative values, an unsigned representation may provide a simpler implementation for some functions. Rather than reserving a bit for a sign (as in sign magnitude) or reserving half of the range for negative numbers (as in two's complement), an unsigned 8 bit field uses all 8 bits to represent a positive value between 0 and 255.

Addition for signed and unsigned values is the same; there is no difference between an adder for "signed" versus "unsigned" values. There is a difference, however, in how the inputs are sign extended, and how overflow is detected.

A signed multiplier differs lightly from an unsigned multiplier. Signed and unsigned comparators also differ.

Saturation check after an unsigned operation is simpler, since saturation processing need not consider a negative result.

Sign extension of unsigned values is simpler, since this is always accomplished by zero padding on the left.

When subtracting two unsigned values, the algorithm designer must determine if the result is signed or unsigned. The difference between two unsigned values can be negative. However, if negative values do not make sense within an algorithm, then an unsigned subtract operation can detect a negative result and force the answer to 0.

two complement values A – B, rather than negating B it can be inferred that, the Timestamp A can be performed by setting the ... complement to the full adder. Extra complement to the last significant ... results in ... the ... binary twos complement subtractor.

Signed Arithmetic

A number can take on only non-negative values in unsigned. In practice often there is a need for... representation for some applications. Rather than see enough ... a way to take a unsigned number setting asides for negative numbers representation conventions where the leftmost field bit of a number to represent sign have calls between 0's and 1's...

Addition, signed and unsigned arithmetic is the same, there is no difference between an unsigned, and signed arithmetic alone. There is a difference between in how the layout here in ... and multiply are handled.

A ... subtracter generates results from an input ... combination. Signed and unsigned subtraction.

Functional construction of ... operation is simpler since saturation occurs and does not affect the result.

A ... overflow of the sign value implies when a result ... exceeds available... in a signed integer ...

When subtracting two unsigned values, the difference ... can be used to determine if the result is positive or negative. The little added between ... and ... result becomes ... If ... number ... register value than the other, then will be ... negative number results in ... of the result group and into the number...

202

Answers To
The Exercises

Appendix

C

Appendix C - Answers to the Exercises

Chapter 1

1. A system requires a signal to noise ratio of at least 48 dB. What is the minimum number of bits required to represent a sample?

    Each bit of a sample represents approximately 6dB of power. Therefore, the minimum number of bits is 48/6 = 8. A/D converters usually have some noise in the least significant bit, so if the input is an A/D converter, one may need 9 bit samples in order to guarantee an SNR of 48 dB.

2. An algorithm computes the sum of 12 separate 8-bit, two's complement samples. In order to represent the sum of all samples with no loss of data or overflow, the sum must be at least what width, in bits?

    Every addition of two samples results in a sample having double the range. If the 12 samples of 8 bits are arranged in pairs, and each pair is added, the result is 6 samples requiring 9 bits. These 6 are then arranged in pairs and added to give 3 samples, each 10 bits wide. Two of the three can be added, resulting in a single 11 bit sample which must then be added to a 10 bit sample. The result will require more than 11 bits, but never reach the full range that can be represented by 12 bits.

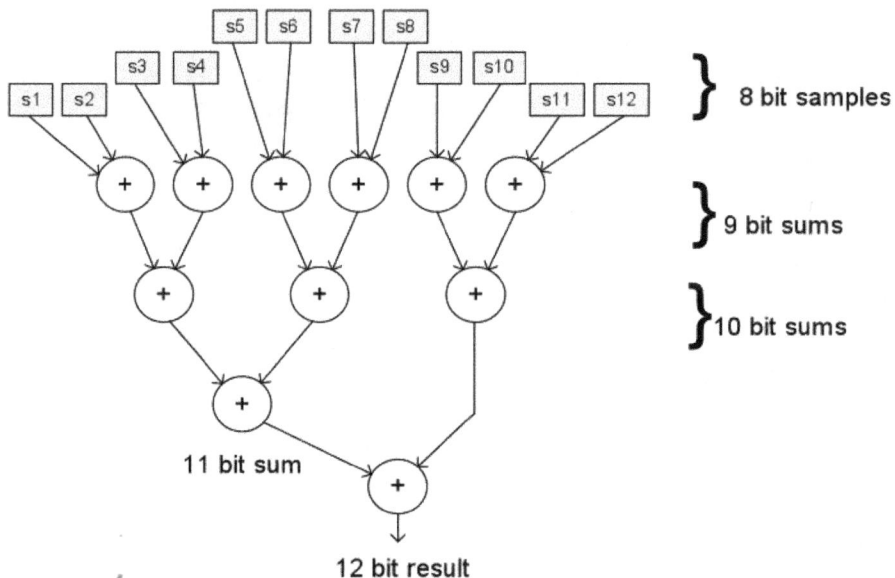

The result therefore requires 12 bits. This is a growth of 4 additional bits beyond the 8 bit samples: $log_2(12) = 3.58$.

Another way to analyze this is to recognize that 8 bit samples can take on the range of -128 to +127. Summing 12 of these will produce a result in the range of -1536 to +1524, for a total range of 3060. The $\log_2(3060)$ is 11.58, indicating 12 bits are required.

3. Re-answer the above question when the values are unsigned 8-bit values.

   This also requires 12 bits. Each 8 bit sample can take on a range of 0 to 255, so the sum of 12 samples will be in the range of 0 to 3060.

4. Suppose we generate an adder for two 8-bit values without any concern for overflow. What does the result look like if overflow occurs?

   When signed addition overflows, the result "wraps" so that adding two large positive values gives a negative result, and adding two large negative values gives a positive result. The actual result is the expected result -2^{nbits}.

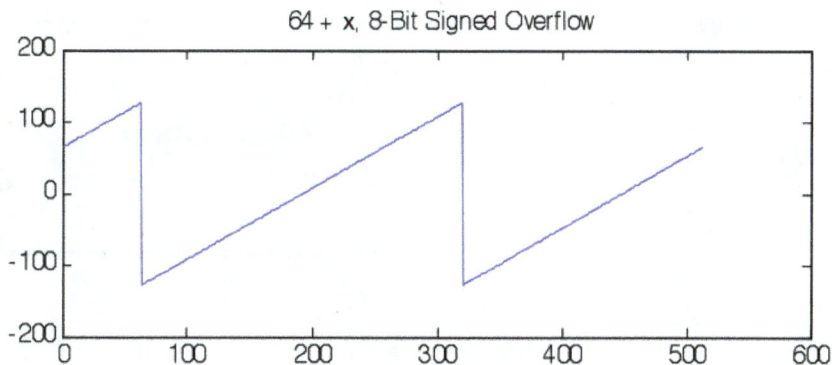

64 + x, 8-Bit Signed Overflow

When unsigned addition overflows, the result "wraps" so that adding two large positive values becomes a small positive value. The actual result is the expected result -2^{nbits}.

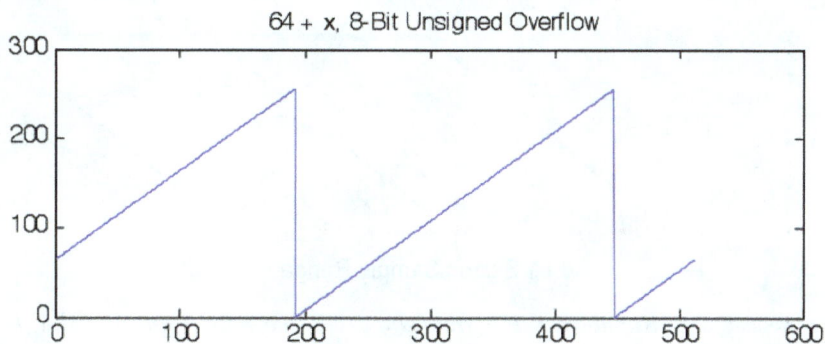

64 + x, 8-Bit Unsigned Overflow

From a signal processing point of view, overflow introduces large excursions or transients into a signal; such transients are wide-band noise and can greatly compromise the performance of a system.

Chapter 2

1. An analog to digital converter converts voltages from -1v to +1v to an 8 bit value where -1v results in a value of 0, 0v results in a value of 128, and +1v results in a value of 255. The algorithms require values of -128 for -1v, and +127 for +1v. What is the minimum amount of logic needed to convert the values?

This is not uncommon. ADCs do not necessarily provide a two's complement output for positive or negative input voltages. Some provide a sign-magnitude output; in this example the ADC essentially provides an unsigned output after biasing the input voltage up.

In this example, the result is easily converted to 2's complement by subtracting 128 from each sample. Thus, -1V becomes -128, 0V becomes 0, and 1V becomes 127.

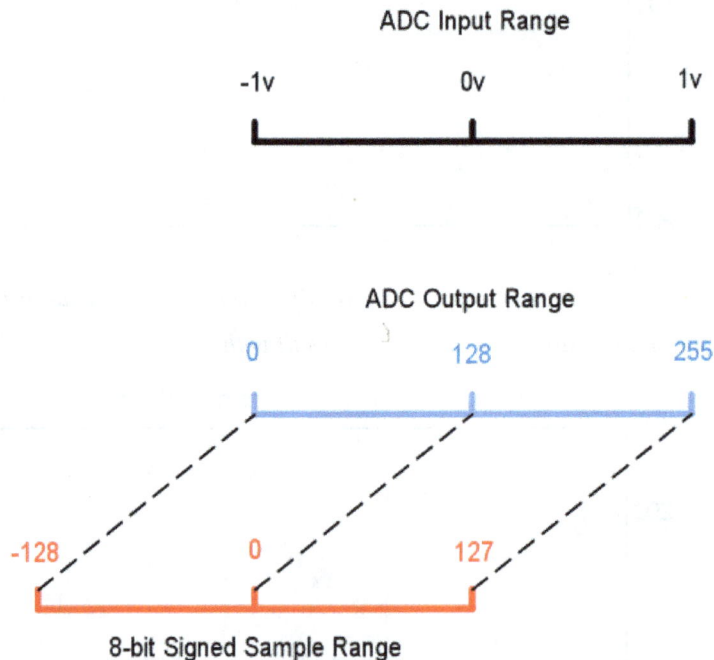

ADC Input Range

-1v   0v   1v

ADC Output Range

0   128   255

-128   0   127

8-bit Signed Sample Range

The easiest way to subtract 128 from an 8 bit two's complement value is to invert the most significant bit. A single inverter on the MSB is the minimum amount of logic required.

Input Value		High Bit Inverted	
Binary	Decimal	Binary	Decimal
0000_0000	0	1000_0000	-128
0000_0001	1	1000_0001	-127
0000_0010	2	1000_0010	-126
. . .			
1000_0000	128	0000_0000	0
. . .			
1111_1110	254	0111_1110	126
1111_1111	255	0111_1111	127

2. Describe the differences between an adder for unsigned values, and an adder for signed values.

There are no differences in the adder itself.

If the adder is coupled with additional logic that performs rounding and/or saturation, then the rounding and saturation logic will differ based on signed or unsigned processing.

3. An 8 bit unsigned value is squared. How many bits wide is the result?

Squaring an 8 bit unsigned number results in a 16 bit number: the maximum input value is 255, the maximum result is 65,025.

4. An 8 bit signed value is squared. How many bits wide is the result?

Squaring an 8 bit signed value results in a 16 bit underlined unsigned number: the input range is -128 to 127; the maximum result occurs when the input is -128; in this case the result is 32768. Note that 32768 cannot be represented in a 16 bit signed number; if a 16 bit signed result is required then the result should be saturated to 32767, or the result should be zero-padded on the left with a single bit to make it a 17 bit signed result. (Zero padding is acceptable here, since the result of squaring is always non-negative.)

5. An FPGA functional block computes the square root of a 12 bit integer. How wide is the result if it is an integer?

A 6 bit unsigned integer result will suffice. The maximum input value is 4095; the square root of this is 63.99. If the result is to be rounded to the nearest integer, then either saturation to 63 should be employed during rounding, or another output bit should be provided, allowing representation of the value 64.

6. An FPGA functional block computes the square root of a 12 bit integer. How many bits to the right of the radix point must be maintained in order to generate the original value after squaring the result?

    In order to obtain a result that, when squared, equals the input to the square root, the number of bits required is 12, where 6 are to the left of the radix point, and 6 bits are fractional. The result, when squared, will equal the input to the square root with a tolerance of +/- 1.

7. When a square root functional block is provided with a negative input value, what output value should it provide?

    This is a system design question. If the system can be structured so that the input is always unsigned, then the square root functional block need not handle this case at all. Otherwise, the designer must determine if the negative input represents an error, or whether there is some value in assessing the magnitude of the input, even though the sign is "wrong". In some designs it may make sense to return 0; in others it may make sense to compute the square root of the absolute value of the input.

    The question cannot be answered without understanding the input constraints, and the effects of forcing the output to a particular value when a negative input is presented.

Chapter 3

1. The signals forming a dial tone are 350 Hz and 440 Hz. Other touch-tone signals also consist of two frequencies, which are not harmonically related, chosen from the table below (ITU [5]). What is the advantage to choosing signals that are not harmonically related?

	1209 Hz	1336 Hz	1477 Hz	1633 Hz
697 Hz	1	2	3	A
770 Hz	4	5	6	B
852 Hz	7	8	9	C
941 Hz	*	0	#	D

Table 5. DTMF ("Touch Tone") Frequencies

Because the signals are not harmonically related, the distortion of any row frequency will result in multiples of that frequency that are not near any column frequencies. For example, 2×697 = 1394, and 3×697 = 2091. The 1394 signal is almost halfway between the 1336 and 1477 columns, so 1394 is not likely to generate a false alarm for a column frequency.

2. *A DTMF detector* is a circuit or a DSP algorithm that detects one of the touch-tone buttons above from the frequencies present within an input signal. Some wide-band signals, such as voices, modem signals, campaign speeches, or elevator music will have touch-tone frequencies present within them, such as 1209 Hz and 852 Hz (a '4' key). Describe a method that allows a detector to detect the DTMF signals, but reject wideband signals that contain DTMF frequency components.

   Such a detector can compare the amount of energy within the row and column with the total energy of the signal: if there is significant energy at exactly one row frequency, and also significant energy at exactly one column frequency, and if the total energy in the row and column constitutes more than, for example, 70% of the total signal energy, then the tones are less likely to be voices or other non-DTMF signals.

3. A perfect A/D converter provides samples of 12 bits of unsigned integer values. What is the signal to noise ratio due to quantization? What is the SNR due to quantization error if the A/D converter performs signed conversion?

   The SNR in either case is approximately 12×6 = 72dB. The device is unable to resolve signals in between the discrete levels that it can accommodate, so the resulting samples will be the nearest 12 bit integer. Discarding (through truncating or rounding) the fractional bits introduces noise (in this case, a small difference between the samples and the actual input signal).

4. Describe the spectrum display for a touch-tone '6' signal.

   The spectrum will have prominent peaks at 770Hz and 1477Hz. Various filter circuits and amplifiers in a telephone or audio system may "color" the signal, so that even if these signals started out with the same amplitude, the receiver of the signals sees the components have different magnitudes. This is referred to as the "twist" of the DTMF signal. Some amount of twist is acceptable.

5. A 770 Hz signal is subtracted from a 1477 Hz signal. Both have equal amplitudes. Describe the resulting spectrum.

   The spectrum will have prominent peaks at 770 and 1477Hz. Subtraction of two different frequencies does not change the magnitude of the spectral components. If one were to examine the phase of the signals, and compare it with the phase of a signal where 770 and 1477 are added together, one would see a 180 degree phase difference in the 770Hz component when comparing the two signals.

6. A circuit is fed a 1000 Hz sine wave, and a 1.310 MHz sine wave, both having amplitude of 1.0 volts peak to peak. The output voltage of the circuit is described by the equation:

$$F(t) = A\cos(2\pi \times 1.310 \times 10^6 t)[1.0 + 0.25 \times \cos(2\pi \times 1000t)],$$

where A is 10.0, and t is time, in seconds.

    a. What is the peak to peak voltage of the output?

    b. Describe the spectrum of the output

    c. Is this circuit a linear circuit? Explain.

This is an AM radio signal, where the carrier is at 1.31 MHz and the audio signal is 1000 Hz.

a. The voltage will be as high as $A \times 1.25$; the peak-to-peak voltage will be twice this, or 2.5A.

b. The spectrum will show three peaks: the carrier, at 1.310 MHz, an upper sideband at 1.311 KHz, and a lower sideband at 1.309 KHz. The carrier will have amplitude A, while each sideband will have amplitude of 0.125A.

c. The multiplication of two sinusoids having different frequencies is not a linear process. A nonlinear process generates new frequencies that were not present in the original input signals. The new frequencies are 1.311 and 1.309.

7. Estimate the SNR of the spectra shown in Figure 9, Figure 10, and Figure 11.

Figure 9: approximately 70 dB. The noise is slightly below 0 dB. A mid-quality audio system might deliver such an SNR.

Figure 10: The clipping introduced multiple noise spikes, some are as high as 40 dB. The noise is computed from the sum of all non-signal spikes; the estimated SNR is about 20 dB. This is a poor quality signal. If voice, it would still be intelligible, although it would sound very noisy, with more noise corresponding to louder portions of the signal. Quieter portions may not clip at all, and sounds reasonably noise-free.

Figure 11: The nonlinear distortion introduced noise in the form of harmonics. Though the output signal has approximately the same shape as the input signal, the harmonics would be very audible and cause the tone to sound excessively "tinny" or "metallic". An SNR estimate for this signal is 10 to 15 dB.

Chapter 4

1. Re-order these complex values from smallest to largest: (1 + 2i), (1 – 2i), 1.0, -7

   The magnitudes of the values, in the order presented, are √5, √5, 1, and 7. From smallest to largest, the complex values are 1, (1+2i), (1-2i), 7. Because the two middle ones have the same magnitude, the result 1, (1-2i), (1+2i), 7 is also acceptable.

2. Compute the sum difference, and product of (1+2i), (3-i).

   Sum:　　　　　　*(1+2i) + (3-i) = (4+i)*

   Difference:　　　*(1+2i) – (3-i) = (-2+3i)*

3. Raise the following complex values to the 8th power: 1, -1, i, -i, (1+ i)/√2.

   All result in a value of 1. The first two are obvious: 1 and 1.

   For both i and -i:

   $$i^8 = (-1)^4 = 1$$

   $$(-i)^8 = (-1)^4 = 1$$

   For (1+i)/√2:

   $$[(1+ i)/√2]^8 = [(1 + 2i – 1)/2]^4 = i^4 = (-1)^2 = 1$$

   These values are 5 of the 8 values which are the 8th roots of 1. The other three are (1- i)/√2, (-1+ i)/√2, and (-1- i)/√2. When plotted in the complex plane, these form a ring of points on the unit circle, evenly spaced around the perimeter of the circle, starting with the real point at 1,0.

4. Geometrically, what happens when a complex value z is multiplied by its conjugate?

   The conjugate is the reflection about the x axis, so if z forms an angle of θ with the x axis, then z conjugate forms an angle of –θ, and has the same magnitude as z. When multiplying two complex values, the result has an angle equal to the sum of the angles (so, θ-θ = 0), and a magnitude equal to the product of the magnitudes. The result lies on the x axis, at a distance equal to the square of the magnitude of z.

5. When any number is multiplied by 1, the result is the original number. What happens when any number is multiplied by a complex number having a magnitude of 1?

   The angles add, and the magnitudes multiply. Since we are multiplying by a number having magnitude of 1, the result has the same magnitude as the original term. However, the angle will be rotated about the origin by an amount corresponding to the angle of the complex term having magnitude 1.

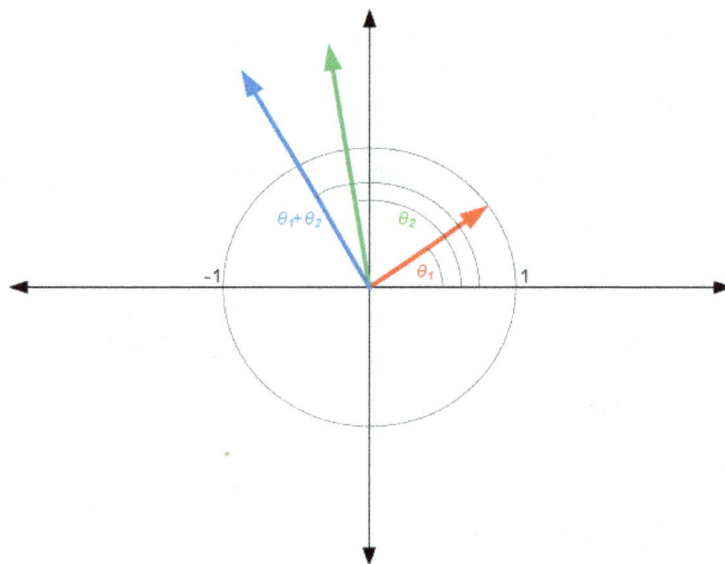

6. What might it mean to raise *e* to a complex value? What might it mean to take the natural log of a complex number?

All functions that can be performed with real numbers can also be performed with complex numbers.

To evaluate e^z, rewrite z as x + iy. Then:

$$e^{x+iy} = e^x e^{iy}$$

Then, note that e raised to an imaginary number iy gives the complex number [cos(y) + isin(y)]

The result is the complex number:

$$e^z = e^x[cos(y) + isin(y)]$$

Note that [cos(y) + isin(y)] is a complex number having magnitude of 1.0 and angle of atan (y/x), and e^x is always a real number. Here the atan function must place the resulting angle for y/x in the proper quadrant; C programmers will recognize this as the atan2(y,x) function. The value e^x determines the magnitude of the result, while y determines the angle of the result within the complex plane.

Note also that if $y = 0$, then

$$e^z = e^x[cos(y) + isin(y)] = e^x[cos(0) + isin(0)] = e^x$$

as we would expect.

The natural log is the inverse operation of e^z; therefore $log(z)$ gives a complex number where the real part is the natural log of the magnitude of z, and the complex part is the angle $atan(y/x)$ of z:

$$log(z) = log(x+iy) = [log(x) + iatan(y/x)]$$

Raising e to the power $[log(x) + iatan(y/x)]$ yields z.

7. Place two stereo speakers so that they face each other. Play your favorite music. Reverse the leads into one speaker. Is the resulting sound different? Explain.

    When the leads on one speaker are reversed, the phase of the sound from the speaker is inverted. This means one speaker "pushes" while the other "pulls". This will be particularly noticeable for loud bass notes, which sound quieter when the phase of one of the speakers is reversed.

Chapter 5

1. Compute the dot product of the vector (1, 2, 3) with

    a. (1, 2, 3)

    b. (2, 3, 1)

    c. (3, 1, 2)

    The vectors a, b, c all have the same magnitude, but different directions. This illustrates the fact that the dot product is maximal when the two vectors being multiplied lie in the same direction:

    a. $(1,2,3)\cdot(1,2,3) = (1 + 4 + 9) = 14$

    b. $(2,3,1)\cdot(1,2,3) = (2 + 6 + 3) = 11$

    c. $(3,1,2)\cdot(1,2,3) = (3 + 2 + 6) = 11$

2. Compute the dot product of the vector (1, 2, 3) with

    d. (2, 4, 6)

    e. (4, 6, 2)

    f. (6, 2, 4)

These vectors have twice the magnitude of the vectors in problem 1. This illustrates the fact that when a vector is multiplied by a constant (2 in this case), the dot product of the vector with another vector is also multiplied by the same constant:

a. $(2,4,6)\cdot(1,2,3) = (2 + 8 + 18) = 28$

b. $(4,6,2)\cdot(1,2,3) = (4 + 12 + 6) = 22$

c. $(6,2,4)\cdot(1,2,3) = (6 + 4 + 12) = 22$

3. Compute the dot product of each vector in exercise 2 with itself.

These vectors all have the same length; we therefore expect the dot product of the vector with itself to have the same result.

a. $(2,4,6)\cdot(2,4,6) = (4 + 16 + 36) = 56$

b. $(4,6,2)\cdot(4,6,2) = (16 + 36 + 4) = 56$

c. $(6,2,4)\cdot(6,2,4) = (36 + 4 + 16) = 56$

4. A single cycle of a cosine, sampled 8 times, has samples:

$$\{1, \quad 0.7071, \quad 0, \quad -0.7071, \quad -1, \quad -0.7071, \quad 0, \quad 0.7071\}.$$

A cosine having twice the frequency has samples:

$$\{1, 0, -1, 0, 1, 0, -1, 0\}.$$

Compute the dot product of these two vectors.

The dot product is $(1 + 0 + 0 + 0 - 1 + 0 + 0 + 0) = 0$; the two vectors represent cosines of different frequencies having an integer number of cycles within 8 samples.

5. A cosine of frequency 0 Hz has samples $\{1,1,1,1,1,1,1,1\}$. Compute the dot product of this with the two vectors provided in exercise 4.

For the first vector, all terms in the dot product cancel. The first vector is a cosine having exactly two cycles, while $\{1,1,1,1,1,1,1,1\}$ is a DC value (0 Hz).

For the second vector, all terms also cancel. Again, this is a consequence of taking the dot product of two different frequencies where the number of samples is able to represent an integer number of cycles for both frequencies.

6. When might one prefer to use the median value of a set of samples instead of the mean value?

If a set of input samples has outliers (extreme values due to measurement errors, or perhaps due to the nature of the physical process), then the mean value may be "pushed" by the presence of the extreme values. If the outliers are few in number, then the median value may be pushed, but only slightly.

7. The mean value of a set of voltage samples is high, while the variance is very low, but nonzero. What does this say about the signal?

   The signal represented by the samples has a prominent DC component, with a small noise or other AC components riding along.

8. The variance of a set of voltage samples is high, but the mean is low. What does this say about the signal?

   The signal represented by the samples has high levels of AC or noise content. A nonzero but small mean indicates it also has a small DC component.

9. A signal has a high mean value and a high variance. What does the signal look like?

   Such a signal has a large DC component and large AC components or noise.

10. Two signals have the same mean value and the same variance. Are the signals alike? Explain.

    The mean and variance are measurements that summarize a signal; there is insufficient information in the statement to determine if the signals are alike. Both signals have the same DC component (mean), but with respect to AC content, unless they both have zero variance, the signals can be very different.

Chapter 6

1. Describe how to use correlation to determine if an input signal contains a dial tone (350, 440 Hz).

   - *Assuming the sample rate is fixed and already determined, choose a number of samples that represent the duration of time over which the process should check for a dial tone. If the number of samples is too small, the correlation will be subject to false alarms, being unable to discriminate between frequencies that are close to 350 and 440. If it is too long, then short dial tone pulses will go undetected. In this case, a recommendation is to start with enough samples to represent 100 to 200 mS. Choose 100 for convenience.*

- *Now, determine the nearest number of samples needed to represent 100 mS, and to represent an integer number of cycles for 350 Hz. Fortunately, 100 mS will represent exactly 35 cycles for 350 Hz. Also, it will represent 44 cycles of 440 Hz.*

- *Correlate the 100mS segment with a complex sinusoid of 350 Hz. This will yield a single complex number. Compute the magnitude of this number.*

- *Also, correlate the 100mS segment with a complex sinusoid of 440 Hz; compute the magnitude.*

- *Compute the magnitude of the signal over the 100 mS period.*

- *Check for signal levels above threshold:*

    i. *The magnitude of the signal should be greater than some minimum value. This eliminates the possibility of low-level background noise, which contains all frequencies, from triggering the detector.*

    ii. *The % of signal energy represented by the 350 and 440 signals should be above some threshold. This prevents high-level broadband noise from triggering the detector.*

    iii. *The ratio of energy in the 350 vs 440 Hz signals should be within an acceptable range. This prevents a false alarm when the amplitude of one of the components is excessive while the amplitude of the other one is negligible.*

- *Finally, note that the input signal data stream, though processed in 100 mS blocks, will not have dial tones that start and stop on 100 mS boundaries. It may therefore be desirable to apply this processing to segments of the input signal stream every 25 mS or 50 mS, so that there is overlap.*

2. What might you expect if a random set of samples is correlated against a random signal?

    If the signals are truly "random noise", then a plot of the correlation will also look like "random noise." Statistically speaking, there are more precise definitions of "random" that qualify this statement.

3. The *autocorrelation* of a signal is the result of correlating a signal with itself. Explain what might cause peak values in the autocorrelation result.

    If a signal is periodic or has periodic components, then shifted versions of the signal will correlate with the original signal. Peaks will occur in the correlation when the shift amount corresponds to one or more periods of the input signals.

If the signal contains no periodic components, then the only peak present in the result will be when the input signal is not shifted at all.

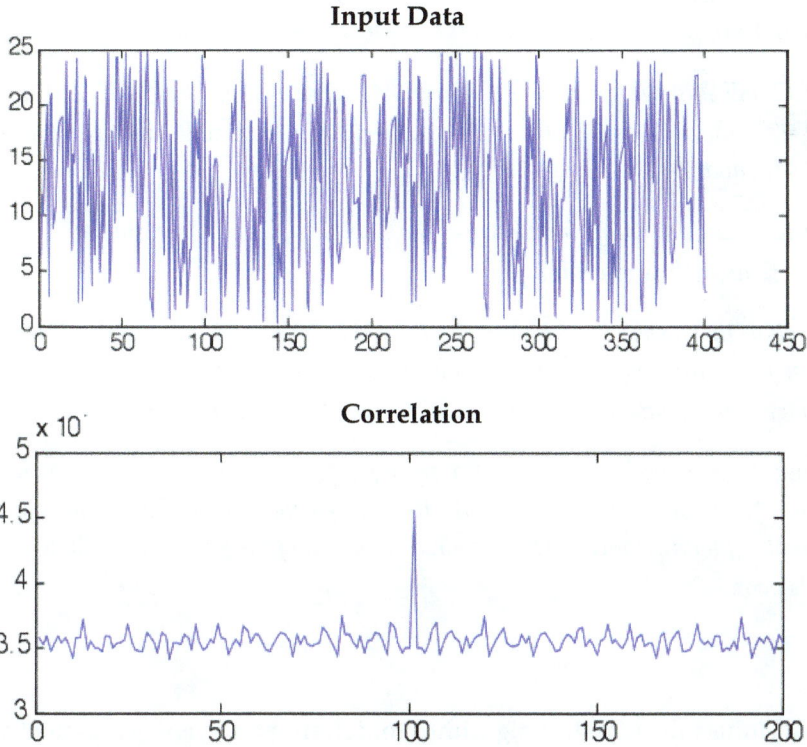

Input Data

Correlation

Chapter 7

1. Describe a plot of the magnitude of the FFT of the point-by-point product of $e^{2\pi j10t}$ and $\cos(2\pi t)$. How many frequencies are present?

    The cosine will generate two peaks: +1 and -1. This is because the cosine is really the sum of two complex sinusoids:

    $$\cos(x) = e^{jx}/2 + e^{-jx}/2$$

    Multiplication of the cosine by a complex sinusoid shifts the center frequency (i.e.,"center tunes" the signal) from 0 to 10, so the two peaks now occur at 11 and 9.

2. Describe a plot of the magnitude of the FFT of the point-by-point product of $e^{2\pi j10t}$ and $e^{2\pi jt}$. How many frequencies are present?

    This plot will show a single frequency at 11. This can be seen in the math:

    $$e^{2\pi j10t} \cdot e^{2\pi jt} = e^{2\pi j10t + 2\pi jt} = e^{2\pi j11t}$$

217

3. A poorly designed NCO produces frequency -f1 at 0 dB, and a spur at frequency -f2 and level -30 dB. This is mixed with a signal having a level of 12 dB at f1, and a stronger signal at f2 having a signal level of 28 dB. The result is low-pass filtered using a filter having 0 dB of attenuation around 0 Hz. Describe the signal content of the result.

The NCO will shift the spectrum by –f1, so that the new signal has 12dB at 0 Hz, and -30dB at f2-f1. The -f2 spur will also shift the signal; the resulting spurs will be added to the output signal: f1-f2 at -28dB, and f2-f2 = 0Hz at -2dB.

No matter what the cutoff frequency of the low pass filter is, the spur at –f2 will mix with the signal at f2 to produce a result at 0 Hz.

Presumably, the system wanted to center tune f1 and filter out all other signals. However, the signal at f2 is large, and it mixes with a weak spur in the NCO at frequency –f2, resulting in a 2dB image of the signal, added to the 12dB image of the signal we want.

The spur, though weak, mixed with a strong signal and produced detectable noise within the passband. Note that in the input, f2 is16dB stronger than f1. In the output, the center-tuned f2 is 14dB weaker than f2. Since f1 is the desired signal and f2 is noise, the SNR is no better than 14dB, which is poor.

Chapter 8

1. We wish to use the Goertzel algorithm to determine the energy content within an audio signal sampled at 8 KHz. The energy content must be evaluated every 50 mSec.

   a. What length should be used for the Goertzel algorithm?

   100 mS is represented by 400 samples at this sample rate. The Goertzel length should be the value closest to 400 which represents an integer number of cycles of the frequency of interest when the frequency of interest is sampled at 8 KHz. For example, a 250Hz signal requires 8000/250 = 40 samples, so 400 is an acceptable length. A 150Hz signal requires 53.3 samples per cycle; 7.5 cycles fit into 400 samples, so a length of 7 cycles (373 samples) is appropriate.

   b. If we wish to compute the complex sinusoid "on the fly", what complex multiplier should we use?

   The complex multiplier will be z = [cos (2π/r) - isin(2π/r)], where r is the number of samples required to represent the specified frequency. For 250 Hz, r = 40; for 150 Hz, r = 1/53.3.

2. A system needs to determine the energy of a signal at 8 specific frequencies over a vector length of 1024 samples. Should the system used Goertzel algorithms, or an FFT?

Assume that the appropriate Goertzel length is 1024 samples. Each Goertzel transform will then require 1024 complex multiply/add operations, equivalent to 4K real multiplies and 4K real adds per Goertzel transform, for a total of 32K real multiplies and 32K real adds.

The FFT requires log2(len) stages of length 1024; each stage requires 512 "butterfly" operations consisting of a complex add, complex subtract, and complex multiply. The total operations are therefore (10 ×512 ×(6 adds + 4 multiplies)) = 5120 ×(6 adds + 4 multiplies) = 30K adds and 20K multiplies. Bit reversal of the output is not needed, since the algorithm can look in the 8 specific bins representing the frequencies of interest.

A trade-off is that the FFT length will always be 1024, meaning that frequencies may be spread across two adjacent bins when an integral number of cycles cannot be represented by 2^n samples.

In this case, the FFT provides more information using less total operations. If the input signal is real instead of complex, then both the Goertzel and FFT operations can be reduced.

3. What might you expect as the output of an FFT from a random set of samples?

   A "random" set of samples contains "random" frequencies, so the FFT and the time domain plot of the signal look like noise. Specifically, if the random set is "Gaussian" and "white", then the spectrum will show noise evenly distributed across the frequency band.

4. What is the resolution (Hz/bin) of a 1K FFT if the input sample rate is 1 megasamples per second? What is the minimum (other than DC) frequency that can be represented? What is the maximum frequency?

   The resolution is 1000000/1024; each bin is 976 Hz wide.

   Frequencies lower than 976 Hz will show some energy in bin 0 (DC), and some in the first bin. If 1K FFTs are performed on consecutive sets of 1024 samples, very low frequencies will show up as undulations in the DC component over a series of sets.

   The highest frequency that can be represented is always ½ the sample rate, regardless of the FFT size. Therefore, the highest frequency is 500 KHz.

5. Describe the FFT of a single pulse having amplitude 1 at time 0, and 0 at all other times.

   The single pulse is called a unit impulse response.

   The real component of all bins will have a value of 1; the imaginary will have a value of 0. This single pulse at time 0 is represented by the sum of cosines having frequencies 0, 1, 2, 3, ... n.

Chapter 9

1. A filter having linear phase introduces a constant delay for all those frequencies which it passes. Why is this phenomenon called linear phase?

   For a given frequency within the passband, the delay, measured in phase angle, is a linear multiple of the frequency. A plot of phase delay as a function of frequency therefore shows a sloped line, which wraps at the 180 degree phase angle.

2. If human hearing is not capable of discerning phase, why would one employ linear phase filtering, when a nonlinear phase filter may require fewer operations per sample?

   Slight changes in phase angle do give an impression of direction and "depth" when heard by both ears, but otherwise phase is difficult to discern. However, signal processing hardware can be designed to change phase, and to detect phase.

   Originally, modems used with telephone equipment varied only the frequency of the modem signal, and were limited to 300 baud or lower. Phase modulation allowed use of higher baud rates, with more sophisticated modulation of amplitude and phase providing up to 56K baud over voice-grade telephone lines.

   Since these modulation methods encode information in the phase, group delays introduced by linear phase filters are acceptable, but non-linear phase filters interfere with information encoded into the phase. Any filters applied to signals employing phase encoding must have linear phase response.

3. A square wave is passed through a linear phase, low-pass filter having a passband of 9 times the frequency of the square wave. Describe the resulting signal.

   The square wave consists of odd harmonics of the fundamental frequency having diminishing amplitude as frequency increases. The more faithfully the high frequency components are reproduced, the "sharper" the corners in the square wave.

   The filter will pass the fundamental, and harmonics 3, 5, 7, and 9.

   The result will be square with rounded corners and slight ringing. Because the filter has linear phase, a group delay will be introduced, so that the output signal will be shifted in time when compared to the input signal.

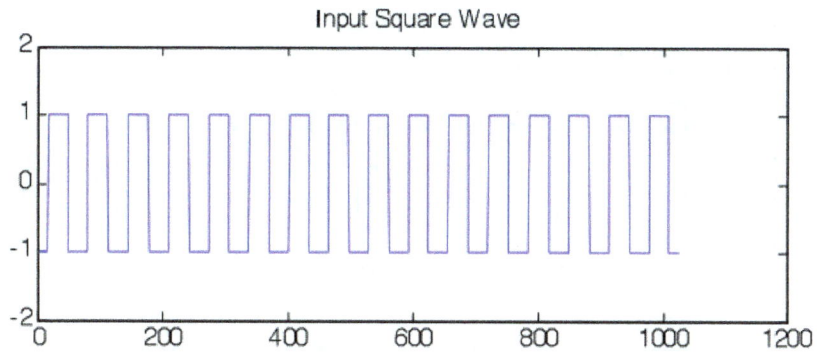

Input Square Wave

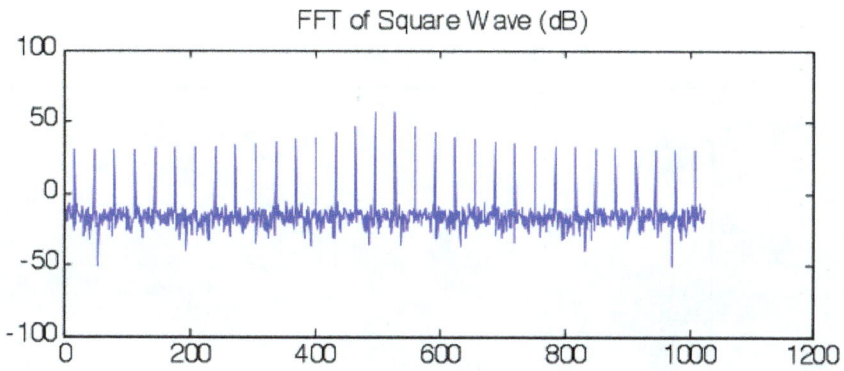

FFT of Square Wave (dB)

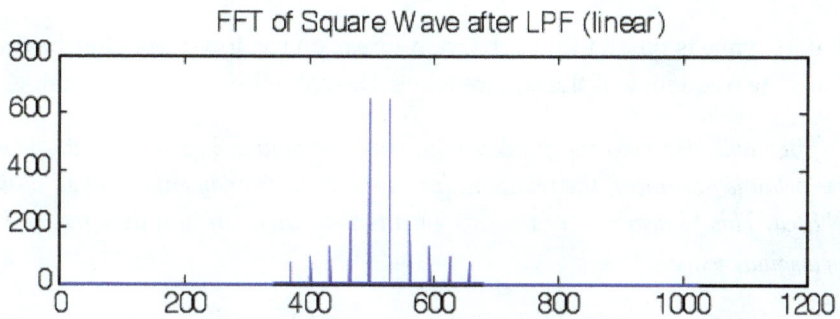

FFT of Square Wave after LPF (linear)

221

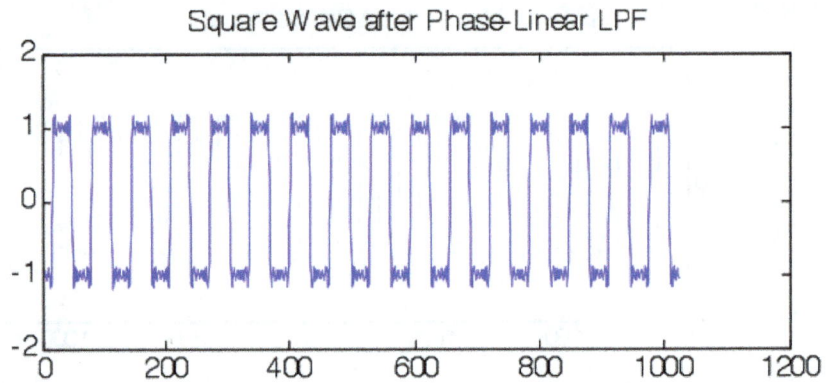

Square Wave after Phase-Linear LPF

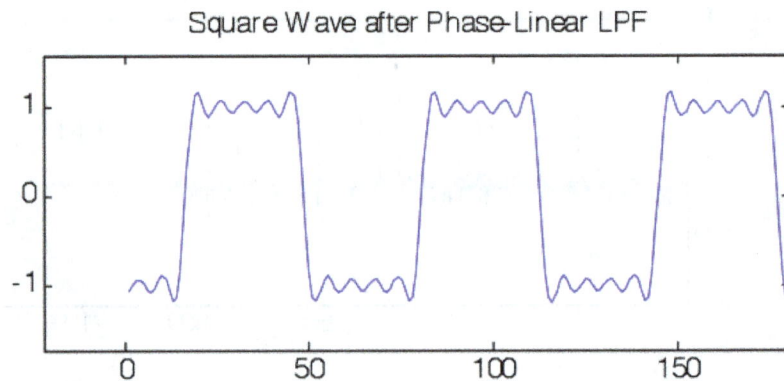

Square Wave after Phase-Linear LPF

4. A square wave is passed through a non-linear phase, low-pass filter having a passband of 9 times the frequency of the square wave. Describe the resulting signal.

This filter will also pass the fundamental, and harmonics 3, 5, 7, and 9. However, because the phase delay is not linear, the relationships between the fundamental and all of the harmonics will be shifted. This "smearing" of the components will cause the output signal to look very different from a square wave.

Interestingly, the signal would sound the same to the human ear as the output signal from exercise 3.

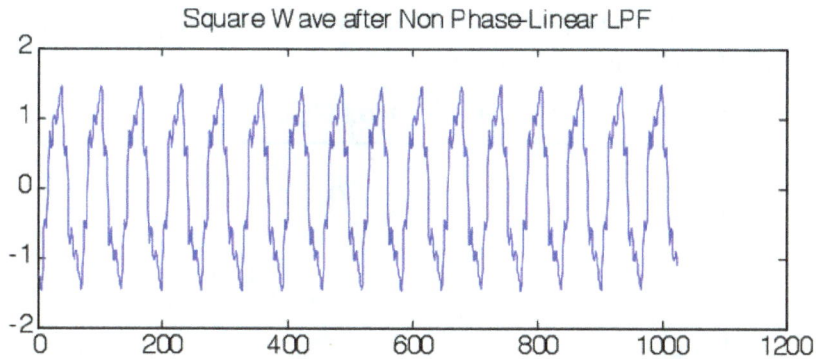

Square Wave after Non Phase-Linear LPF

5. Assume the filters in exercises 3 and 4 have the same passband frequency response.

a. Compare plots of the magnitude spectrum for each filter output.

The magnitude of the spectral components will look the same.

b. Compare plots of the phase spectrum for each filter output.

The signal from exercise 3 will have phase angles that form a straight line. The signal from exercise 4 will not.

Chapter 10

1. Linear interpolation is the process whereby data between two points is estimated by assuming that the two points form a straight line; the estimated point is then proportional to the distance between the points on the X-axis and the slope of the line connecting the points. Since linear interpolation is computationally efficient, why use filtering to accomplish interpolation in the signal processing environment?

The assumption that two adjacent points are connected by a straight line is incorrect when the signal is generated by the sum of sines and cosines. The sine and cosine functions are constantly changing as a function of time, and, because their derivatives are also changing, they never assume a straight line.

Use of straight-line interpolation, though it may provide a result that is pleasing to the eye, therefore does not represent the actual signal. Though it is a form of interpolation, if applied to signals, it introduces noise.

2. A system requires that the frequency components of a set of 200 samples be computed. An engineer proposes that the samples be padded out to a length of 256, so that a radix-2

or radix-4 FFT can be performed. What effect does zero padding the samples have on the magnitude of the FFT?

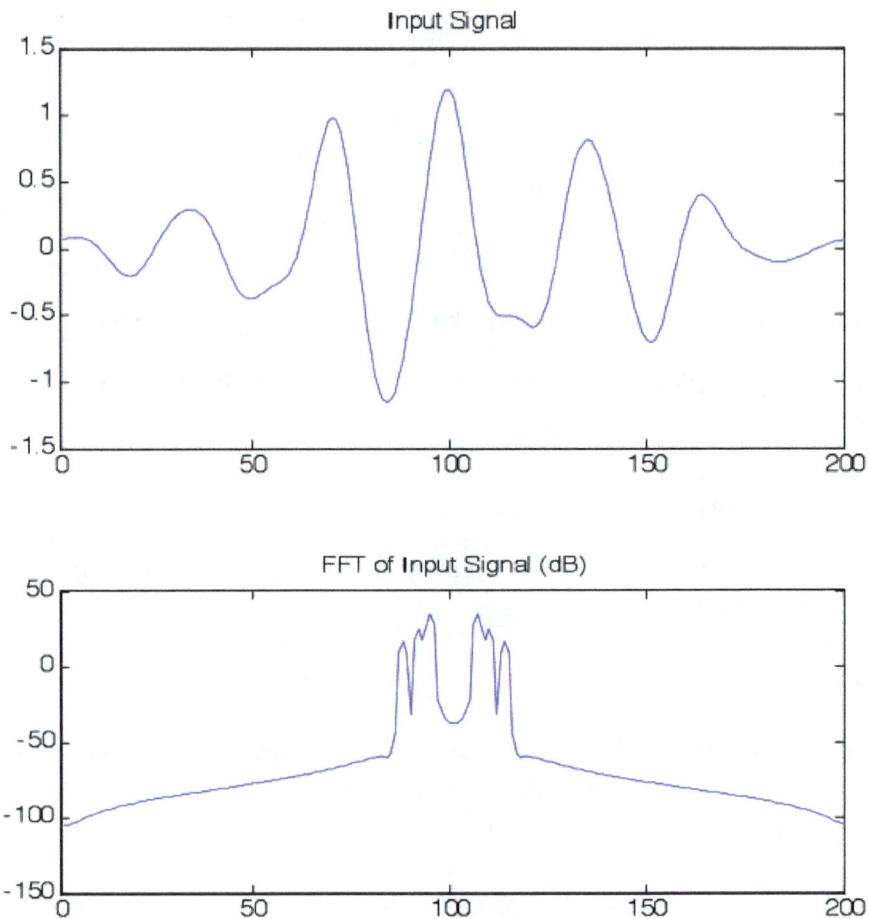

Input Signal

FFT of Input Signal (dB)

Padding with zeroes in the time domain is equivalent to interpolation in the frequency domain. The spectrum spreads out. The energy per bin also decreases, since the padded input signal has many samples having zero energy.

Padded Input Signal

FFT of Padded Input Signal (dB)

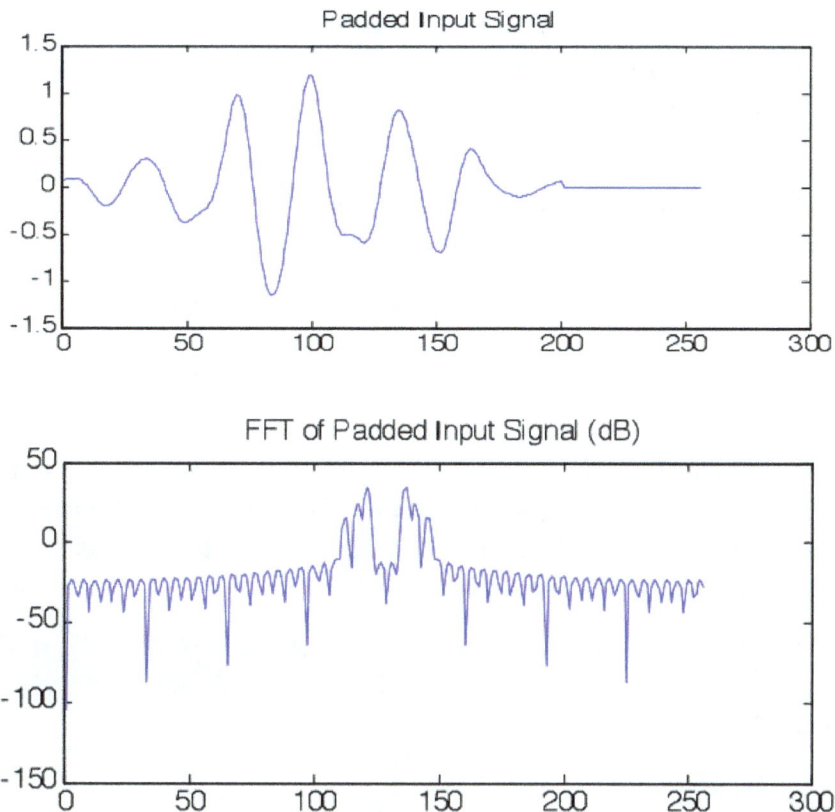

Note that no new information is created: though the FFT has more data points, "new peaks" do not occur. If the zero padding results in a vector that is N times the length of the original vector (where N is an integer), then the FFT will have the exact same shape as the original FFT. Otherwise, the new FFT will look slightly different due to the fact that some energy spreads into adjacent bins.

3. A certain system samples a sensor to produce a sample stream at 250K samples per second. The information of interest lies between 100 KHz and 120 KHz. Propose various approaches for operating on the sample stream that eliminates any signals outside of the 100 KHz to 120 KHz region.

   a) *Generate a bandpass filter which passes 100KHz-120KHz. The resulting sample rate still needs to be 250K samples/sec.*

   b) *Mix the input signal with a complex sinusoid of frequency -100KHz. Low-pass filter the complex result; take the real part only. This will be a real signal in the range 0 to 20KHz. The signal can be decimated by a factor of 5, resulting in a 50K sample/sec real signal.*

c) Mix the input signal with a complex sinusoid of frequency -110KHz. Low-pass filter the complex result; keep the real and complex components. The result is a complex signal in the range -10 to +10KHz. This can be decimated by a factor of 10, resulting in a 25K sample/sec complex signal.

Note that in b) and c), the same information is present, though the representation differs slightly: b provides 50K real samples/sec, while c provides 25K complex samples per second.

4. Does decimation change the signal to noise ratio? If so, how?

The decimation operation itself is noise-free. The low-pass filter, however, can introduce some noise if the coefficients are imprecise (truncated or rounded).

A decimation filter can also result in a reduction in quantization noise present in the input signal: if the quantization noise is random, then the decimation filter can improve SNR because each output sample is the weighted average of N input samples, where N is the filter length. The mean value of the noise remains unchanged, while the signal level increases, resulting in improved SNR.

Index

multiple, 21, 32-34, 36, 39, 44, 47, 69, 106, 114, 116, 120, 126, 140-141, 145, 149, 153-154, 162, 174, 177, 209, 210, 220

multiplication, 9, 17-20, 24, 28, 57-59, 61, 65, 116, 127, 153, 185, 190, 195, 199, 210, 217

multiplier, 4, 17-19, 23, 57, 61, 99, 112-114, 129, 194, 201, 218

music, 36, 42, 45, 51, 67, 209, 213

nature, 69, 186, 215
nbits, 205
NCO, 94, 101-108, 218
ndimensional, 75
negate, 7, 13-14, 16-17, 95
negating, 58, 96, 112, 195, 201
negation, 9, 13-14, 16-17, 58, 197
negligible, 35, 216
neighbor, 159
noise, 1, 4-6, 8, 21, 31, 33-35, 37-39, 43, 46-47, 49, 51, 63, 69, 79-80, 85-86, 106, 121, 123, 125, 127-128, 132, 149, 159, 164-165, 168, 172, 176, 196, 198, 204, 206, 209-210, 215-216, 218-219, 223, 226
noisy, 80-81, 84, 210
Nomenclature, 188
nonlinear, 35, 40-41, 148, 192, 210, 220
nonzero, 21, 27, 59, 77, 104, 111, 137, 142-143, 190, 195, 215
normalization, 25, 76, 190
normalize, -ed, -ing, 25, 76, 82, 84, 177, 181, 189, 190
normally, 4, 35, 60, 62, 85, 100, 102, 110, 123, 134, 136-137, 141-143, 171, 176, 178, 187-188, 191, 194
notation, 2-3, 65, 188
notch, 136, 143
nPoints, 71-72
nSec, 98
ntaps, 139
NTE, 138
nuances, 9, 18

nudging, 104-105
null, 175
numerator, 59
numerical, 1-2, 5, 11, 49, 132, 199
numerically, 101, 198
numerous, 41, 47
Nyquist, 118

occurrence, 79-80
octave, 31, 42-43
oddity, 3
OFL, 195
ohm, 38
operand, 3, 12-13, 15-17, 24, 189-190, 195, 199
operation, 2, 10-11, 14, 15, 18, 53-55, 62, 65, 70, 88-90, 97, 112, 116-117, 128, 140, 148, 150, 156, 186, 189, 195, 197-198, 201, 213, 219-220, 226
operator, 3, 10, 46, 177
Oppenhiem, 117, 129
option, 117, 119, 121
optional, 10, 12, 36, 178
org, 52
origin, 74, 94, 212
originate, 97, 168
origins, 94
orthogonal, 74-75, 87-88, 111, 186
oscillate, 94, 142, 171
oscillator, 61-62, 101, 172, 176
oscilloscope, 36, 110, 186
outliers, 215
outlying, 70
Overdriving, 35

parameter, 18, 32, 49, 62, 70, 94, 131-132, 137-138, 141, 175, 177, 180
parity, 178
Parseval, 114
passband, 49, 136-141, 148, 152-154, 156-157, 160, 188, 218, 220, 222-223
path, 163, 192

stddev, 72, 75, 180
stereo, 67, 213
stopband, 136-137, 139, 141, 157
storage, 2, 103
strategy, 21, 185
stymied, 54
subscript, 89
subset, 70, 161
Subsonic, 31
subtraction, 3, 9, 15-17, 22, 28, 55, 195, 197, 209
subtractor, 15, 153, 194-195
succession, 161
suitable, 176
summarize, 69, 153, 215
summation, 112, 120
superimposed, 164
superimposes, 43
suppress, 43, 172
suppression, 152, 154, 157, 174
susceptible, 98
Swap, 127
symbol, 173-175, 177
symmetric, 46, 89, 123, 128, 135, 139
symmetry, 117, 139-140
synchronize, 178
synthesized, 102, 172
system, 1, 5, 6, 8-10, 31-33, 35-36, 39, 43, 52, 63, 67, 69, 80-81, 89, 129, 132, 141, 163, 165, 168-171, 173, 186-187, 190-191, 204, 206, 208-210, 218, 223, 225

talker, 63
tan, 60
tangent, 25-26, 28, 60
technique, 46, 48, 49, 103, 159
technology, 170
telecommunication, 5, 8, 52
television, 48, 168
temperature, 163
tempted, 138

textbooks, 49
THD, 35
theorem, 79, 88-90
thermal, 34-35, 39
threshold, 24, 50, 73, 172, 174-175, 187, 216
time-varying, 40
tolerance, 172, 208
tone, 34, 37-39, 51, 91, 164, 173-175, 177, 208-210, 215-216
toolkits, 121
top, 38, 39
tradeoff, 44, 188
transaction, 72
transceivers, 46
transcendental, 62
transducers, 36
transform, 74, 98, 109-111, 113-116, 121, 126, 181, 219
transformed, 125
Transformer, 99
transients, 206
translation, 175
transmission, 5, 32, 36, 44, 46, 79-80
transmit, 32, 36, 38, 46
transmitter, 39, 46, 168, 177
triangle, 56
trigonometric, 28, 53
truncate, 18, 22, 190, 197
truncated, 17, 21-23, 49, 185, 197-200, 226
truncating, 5, 21, 23, 185, 194-200, 209
truncation, 5, 9, 13, 15-17, 21-23, 35, 185-186, 195, 197, 199-200
Tsize, 104
TV, 31, 48

UHF, 35
ultra, 170
ultrasonic, 31, 36
ultrasound, 36
unaltered, 23
unambiguous, 97

The Company

The **Amches** Team supplies mission-critical technical expertise and service in support of national security. For nearly a decade, Amches (pronounced *Am-chess*) has provided a broad range of data management services on key contracts with the Department of Defense, and built a reputation as a respected and trusted partner in the Intelligence Community.

During recent years the United States, its citizens, and its business have come under constant threat of cyber-attack. We see it every day in email spam, identity theft, fraud, website phishing, and hacker attacks on business and Government Systems. Amches is dedicated to helping protect our nation's government, people and businesses.

To support national security in a cyber world, enormous amounts of data must be collected, housed, secured, and processed. Once the data has been analyzed, critical information is provided to key decision makers. Amches comprehensively supports this process by providing the following targeted consulting services:

- Database Application Development and Administration
- Software Engineering
- Systems Engineering
- System Administration
- CMMI Processes
- COOP – Mission Assurance
- Data Warehousing
- Business Intelligence
- Project Management Services
- Agile Project Management and Processes

Amches employees are selected for their ability to balance knowledge, autonomy and ethics along with a passionate drive to *"get the job done right."* Drawing on decades of combined experience, Amches is fully prepared to tackle the most difficult government and corporate data management needs.

We provide unique and cutting-edge solutions to the most vexing problems faced by government and commercial entities. Insider Threats, Advanced Persistent Attacks and the "low and slow" exfiltration of critical data are example areas of concern where we are providing competent, cost-effective solutions to demanding clients. We are developing unique techniques and algorithms for protecting wireless point-to-point infrastructure from man-in-the-middle or spoofing threats that can be economically deployed on simple FPGA cards, DSPs or embedded in RFID and Smart Cards. We are also teamed with industry leaders to bring specialized online transactional security solutions to your public-facing web site and provide static and dynamic code analysis for effective protection from existing and emerging vulnerabilities.

Our Services

Systems Engineering for Mission-Critical Applications

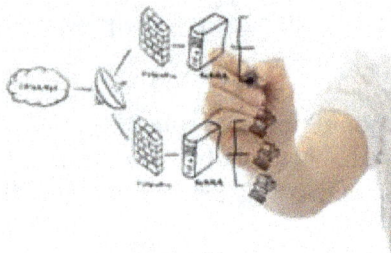

Our clients want their system engineering challenges to be met quickly and accurately. We achieve this by using software project management and lifecycle tools that are familiar to our team.

We're proficient in the following areas:

- Capability Maturity Model Integration (CMMI) process support
- Process Re-Engineering
- Requirements Definition
- Risk Management

Our team is also proficient in the fields of Operations Management Support and Systems Engineering and Technical Assistance (SETA) support.

Rapid Deployment of Software and Database Systems

When software and database problems threaten to derail our clients' business, our employees respond with high end software development technologies.

These technologies include object-oriented design and analysis and rapid software development through the use of the popular Java platform, which also includes Java 2 Platform, Enterprise Edition (J2EE) and Enterprise JavaBeans (EJB).

In addition, the company provides database administration, data and information management, and extract, transform and load (ETL) support.

Amches has also worked closely with the federal government in their ongoing homeland security operations. We have supplied technical expertise in the following areas:

- Database Application Development and Administration
- Data Warehousing
- Systems Engineering
- Continuity of Operations (COOP) Mission Assurance
- Business Intelligence
- Agile Software Engineering Frameworks and Methodologies
- Capability Maturity Model Integration (CMMI) Processes

Contact Us

Contact us at www.Amches.com for more information on any of our products, services, and publications.

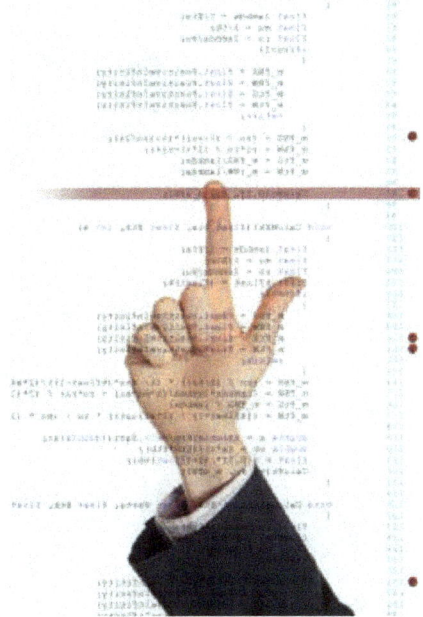

www.ingramcontent.com/pod-product-compliance
Lightning Source LLC
Chambersburg PA
CBHW061354210326
41598CB00035B/5987